EXPLORING HAPPINESS

EXPLORING
HAPPINESS

FROM ARISTOTLE TO BRAIN SCIENCE

SISSELA BOK

YALE UNIVERSITY PRESS
NEW HAVEN AND LONDON

For information about this and other Yale University Press publications, please contact:

U.S. Office: sales.press@yale.edu   www.yalebooks.com
Europe Office: sales @yaleup.co.uk   www.yaleup.co.uk

Set in Arno Pro by IDSUK (DataConnection) Ltd
Printed in the United States of America

Library of Congress Cataloging-in-Publication Data
Bok, Sissela.

  Exploring happiness: from Aristotle to brain science/Sissela Bok.
    p. cm.
  ISBN 978-0-300-13929-7 (cl:alk. paper)
  1. Happiness–History. I. Title.
  BJ1481.B642010
  170–dc22

                                                              2009052121

ISBN 978-0-300-17810-4

A catalogue record for this book is available from the British Library.
10 9 8 7 6 5 4 3 2 1

CONTENTS

I

LUCK

THE MIND REELS AT THE THOUGHT OF THE INFINITESIMAL chances that any one of us had of being born, able to relish even the slightest glimmer of happiness. In my own case, I can pinpoint one of the myriad moments when all such chances would have been eliminated for good. It occurred four years before I was born. Were it not for my young mother's newfangled ideas about happiness, I would never have seen the light of day.

My mother, Alva Myrdal, had come close to death after a traumatic miscarriage, then developed a large uterine tumor. Her doctors had urged her to undergo a hysterectomy to remove the tumor, warning her of the high risks to her life from any future pregnancy. But she had refused point blank to have an operation that would have rendered her sterile. Having studied with the psychologist Alfred Adler, she was convinced that there could be little happiness for parents of lone, often self-centered children in what she called "miniature families."[1] She longed to have siblings for the one child she and my father, Gunnar Myrdal, had at the time – my brother Jan, three years old.

Five decades later, I came across a letter in which my father wrote of his support for her decision to refuse the hysterectomy and instead opt for a more difficult operation to remove only the tumor. "We are putting the outcome in the hand of fate. We have always had magical luck up to now with all that we have undertaken, and I don't understand why that should not be the case again."

As I read that letter, I could see my life, still to come, hanging by a thread, while my mother-to-be, shy and uncertain though she must have felt when confronted with the doctors' expert opinion, insisted on having her own way. Only thanks to the joint decision that my father describes and the luck he and she hoped would continue to be theirs did I come into this astounding world at all. For that opportunity I continue to be awed and humbly grateful. Many describe the same feeling after coming face to face with death, perhaps on learning that they are in remission from cancer or that they have survived a tornado or an airplane crash.

When I look at my life and those of all others as resulting thus from innumerable accidents I see these lives as strange and unpredictable adventures. It makes no sense, from such a perspective, to settle into the rut of blaming parents, society, or fate for the course one's life has taken; or to feel locked into some particular mold that nothing can help one crack open. Yes, we are buffeted by forces and random events far beyond our control. But this is no reason to stop generating efforts of our own, to alter the situation in which we find ourselves, as my mother did. We become who we are in part by how we respond to the shifting circumstances against which our lives delineate themselves.

Happiness, in such a perspective, is to be sought out, pursued, striven for, as in the myths and folk tales about young persons setting out to seek their fortune. They have no assurance of success, no assurance that happiness is owed to them. They have to traverse unknown regions, encounter seductive lures, take highstakes risks, sometimes come back empty-handed. They must find the right balance between empathy and resilience – between fellow-feeling and self-protection – as they learn to perceive the humanity and the urgent needs of many a strange-looking creature, while remaining wary of all who claim to know the one and only path to happiness.

The same is true for anyone embarking, as I do in this book, not so much on a quest to *find* happiness, still less to prescribe steps for others to take to achieve it, as to explore what we can learn about its nature and its role in human lives. Not since antiquity have there been such passionate debates as those taking place today about contending visions of what makes for human happiness. The sixth and fifth centuries BC saw an unparalleled display of such visions by poets, prophets, and thinkers such as Confucius, Buddha, Zoroaster, Lao-Tse, and Socrates, often challenging ordinary perceptions as illusory and describing other, distinctive, paths to true happiness. By the first centuries AD, claims about how to achieve salvation, happiness, and bliss resonated in Eastern cults and mystery religions. They competed not only with Christianity, Judaism, and the worship of Greek and Roman deities but also with secular teachings such as those of Aristotle or Epicurus about earthly happiness as the highest human good.

Now as then, vast populations are on the move, with travelers bringing unfamiliar creeds from distant regions and with migrants, driven by wars and privations, encountering novel visions of the good life. Now as then, the various ideals of happiness carry fundamental moral teachings about how to live. And just as the seekers in myths and folk tales need more than a little luck in order not to return empty-handed, so does anyone examining the role of happiness today. Just as those seekers benefit from combining sympathetic understanding with a dose of healthy skepticism, so too does anyone who ventures into the jungle of claims and counterclaims about happiness – especially when they meet up with conflicting appeals by religious, political, and other authorities to set aside all misgivings and place faith in their dictates.

In setting out to explore the subject of happiness, my aim was twofold. First, I wanted to ask what we can learn about it from bringing together the striking new findings of natural and social scientists with long-standing traditions of reflection by philosophers, religious thinkers, historians, poets, and so many others. Recent decades have witnessed a remarkable growth of scientific studies shedding light on what people say about their experiences of happiness and well-being, on fluctuations in the brain during such experiences, and on the interaction between heredity and the environment. How might we best draw on the new research – linking the past and the present, spanning the various disciplines? Taken together, these different approaches can contribute to a fuller, deeper understanding of the scope of happiness, even as some among them can point to shallowness, tunnel vision, or errors in others. Examining the divergent conceptions of happiness side by side will allow us to fathom their richness and to weigh the clashing arguments about how to capture the full experience of happiness, what contributes to it or detracts from it, how it might be defined and measured, and how it relates to income, say, or to temperament, sociability, marriage, or religious faith.

My second aim was to consider, against this background, perennial moral issues about how we should lead our lives and how we should treat one another. What are the wisest steps to take in the pursuit of happiness? What moral considerations should set limits to such pursuits? What else should matter in human lives aside from happiness? How should we weigh our own happiness against that of others in a world where we are aware, as never before, of extremes of misery and opulence? How might we best take into account what we are learning about the effects of our individual and collective choices on the prospects for the well-being of future generations? And how should we respond to individuals and groups advocating intolerant or outright inhumane conceptions of happiness or well-being?

Bypassing such moral issues makes it all the easier to give short shrift to assumptions that form the subtext to even the most innocuous-seeming views of happiness. These assumptions concern power – power exerted or defended against, whether in families, communities, or political and religious institutions. Often unspoken, these assumptions are about who has the right to pursue happiness, who does and does not deserve happiness, and whether the happiness of some requires the exclusion or exploitation of others. Today, conflicts over them are playing out on a far larger stage than ever before, reaching billions of individuals across the globe, their fortunes affected by global economic shifts beyond their control, their hopes fanned by mass media promotion of methods for achieving happiness in daily life or for finding the path to eternal bliss.

To refocus attention on the moral dimensions of the pursuit of happiness, I ask, throughout this book, what I call "Yes but" questions in the face of claims that a particular action or personality trait or belief or way of life will bring greater happiness. Some of these questions are of an empirical nature, asking for evidence to support such claims or voicing caution in the face of their frequently cheery, upbeat appeal. Others are of a moral nature, asking whether it would be right to seek the happiness held out as desirable or to enjoy it, once it was achieved. Will pursuing such happiness involve us in deceit? Will it require that we break a promise? Is it cruel, unjust, exploitative? Does it call for us to blind ourselves to needs we would otherwise feel duty-bound to address? Stepping back to ask such questions creates space for reflection, for seeking to perceive more fully and to deliberate more attentively in the face of the many conflicting claims about what happiness is and how it should be pursued.

As I was beginning to work on this book, some colleagues in the fields of philosophy and public health brought up an objection to the entire undertaking – one so common that it may be a natural first response to hearing that someone is writing about happiness. Why study that subject now? Is it not a luxury to do so, given the anguish and insecurity of our own time and the number of people who live in dire poverty, devastated by wars, threatened by hunger, epidemics, droughts, and floods? Shouldn't my inquiry begin, rather, by focusing on suffering?

Yet it is precisely in times of high danger and turmoil that concerns about happiness are voiced most strikingly and seen as most indispensable. From earliest times, views of human happiness have been set forth against the background of suffering, poverty, disease, and the inevitability of death. The Roman Stoic thinker Seneca wrote his most moving letters on the subject to his friend Lucilius while being hunted by the henchmen of the Emperor Nero who finally forced him to commit suicide. In sixteenth-century France, Michel de Montaigne conveyed his enduring delight in many aspects of daily life, despite having spent most of his years in the shadow of war and pestilence. And when Thomas Jefferson included in the American Declaration of Independence the inalienable rights to "life, liberty, and the pursuit of happiness," he surely did so at a time of exceptional insecurity and of massive threats to life and liberty. In our own time, consider the juxtaposition of happiness and grim reality conveyed in the statement by Archbishop Desmond Tutu, describing his feelings in April 1994 when he and millions of black South Africans could finally vote:

> The moment for which I had waited so long came and I folded my ballot paper and cast my vote. Wow! I shouted, "Yippee!" It was giddy stuff. It was like falling in love. The sky looked blue and more beautiful. I saw the people in a new light. They were beautiful, they were transfigured. I too was transfigured. It was dreamlike. You were scared someone would rouse you and you would awake to the nightmare that was apartheid's harsh reality.[2]

More recently still, Ingrid Betancourt, when rescued in July 2008 after six years of being kept as a hostage in the jungles of Colombia by the Revolutionary Armed Forces of Colombia (FARC), described her emotions on embracing her children: "Nirvana, paradise – that must be very similar to what I feel at this moment. It was because of them that I kept up my will to get out of that jungle. It's like being born again. If I live 100 years to be an old lady with gray hair, I'll keep marveling at what I saw, what I lived through yesterday. It's a miracle."[3]

The study of happiness never was a luxury to be postponed until more serene, peaceful times. But exploring it may be needed even more in our time, given the unprecedented shift in how most people perceive the possibility of happiness in their own lives. During the course of the twentieth century, societies the world oversaw dramatic reductions in illiteracy, infant mortality, and premature death. By the end of the twentieth century, average life expectancy even in some of the world's poorest societies, such as Bangladesh, was higher than Britain's at the beginning of that century. Most of the world's peoples now enjoy standards of living and social and political freedoms unimaginable to their great-grandparents. Women have far more opportunities than they had in the past; and even in societies where they are most severely held back, mistreated, and exploited, increasing numbers are learning that this need not be their condition.

At the dawn of the twenty-first century, therefore, ancient notions about the need for submissive acceptance of misery and discrimination are losing their power. In many societies, people would scarcely comprehend claims such as that by Samuel Johnson, echoing Ecclesiastes, that no one could wish to be born who knew beforehand all the miseries that would await him in life. They might disagree as fiercely as ever about what happiness is and about what factors make it more or less likely, but far fewer deny that it is at least possible. The levels of suffering and deprivation that beset so many the world over are rightly seen as the more unjust because they are unnecessary, given the vast resources available to overcome them.

I have approached the subject of happiness from several perspectives: by seeking out accounts of the experience of happiness in its own right; by asking how it has been analyzed by philosophers, theologians, and historians; and by considering the rich scientific resources now becoming available in fields such as psychology, economics, health care, genetics, and the brain sciences. Using the first approach, I have looked for accounts of different experiences of happiness and asked what we can learn about such experiences from introspection, personal narratives, thought-experiments, literature, and art. To neglect these deeper, sometimes more intimate forms of testimony is to waste a precious resource for the study of happiness. After all, we respond so much more directly to what Archbishop Tutu and Ingrid Betancourt say about their experiences of happiness than to dictionary definitions or to statistics comparing age groups or nationalities.

By itself, however, this first approach deprives us of the analysis and the empirical findings shown in contemporary studies and of debates about how to define, measure, and investigate the moral issues happiness raises. These approaches allow us to consider how the new findings challenge or confirm the results of introspection and reflection. Without the analysis and information now available, a perennial temptation has been to issue sweeping generalizations about the state of human happiness. Theologians contrasting the miseries of earthly existence to heavenly felicity have been as likely to utter grim estimates on this score as secular thinkers who declare that most people lead lives of quiet desperation. Conversely, both secular and religious visionaries through the ages have proclaimed that happiness is within the reach of everyone willing to accept their particular doctrine and to undertake the changes they prescribe. So have a number of self-help guides to a happier life, again both secular and religious, such as Norman Vincent Peale's 1952 best-seller, *The Power of Positive Thinking*.[4] Long before Peale declared that if you say to yourself that life is good and that you choose happiness, you can be quite certain of having your choice, books such as Horace Fletcher's *Happiness as Found in Forethought Minus Fearthought* were among many that promised happiness to those willing to undertake the particular personal changes the authors recommended.[5] By now, countless books and websites invite readers to learn how to find happiness, become rich, find love, and live longer, healthier lives. Many stress the element of individual choice and control in such a process, as in the scores of books with titles inviting readers to "choose happiness," or "choose to be happy."

Surveys of what people the world over actually say about their own experience contradict both dismal and exultant generalizations. Fortunately for humankind, most people do not see themselves as leading lives of quiet desperation; instead, the majority among them, even in poor societies, regard their lives as moderately or very happy. However, it is equally wrong – and indeed sentimental – to imagine that happiness has nothing to do with standards of living, that it can be achieved equally well by all persons in spite of poverty, ill health, or denials of basic human rights; or to assume that levels of happiness alone should count regardless of concern for such rights. Although some people can be happy even in direst penury, individuals in democratic nations with higher average incomes and standards of living are much more likely to report feeling satisfied with their lives than those who live in the poorest and most oppressive societies. On this point, all empirical studies agree. In fact, researchers have found that economic growth, freedom of choice, respect for human rights, and social tolerance all contribute to greater happiness.[6]

It has given me special pleasure, in preparing this book, to follow different lines of study as if I were in the company of individuals engaged in the same pursuits. Some of their inquiries reach across millennia, as when Teresa of Avila reflected, in her *Life*, on St Augustine's *Confessions* or when Montaigne engaged in imagined exchanges with Socrates, Seneca, Horace, and many others, just as thinkers ever since have done with him. Similar dialogues are taking place today, when neuroscientists, psychologists, and Tibetan Buddhist monks collaborate on studies of consciousness, meditation, and the role of positive and negative emotions; or when economists, sociologists, and psychologists debate the relevance of surveys of life satisfaction in different societies.

Early on, I felt fortunate to encounter books by three authors more attuned to such dialogues than most: the French historian Robert Mauzi's study of the idea of happiness in the eighteenth century, a time when so many thinkers saw the quest for happiness as the great innovation, the glorious discovery of their time; the American philosopher V. J. McGill's *The Idea of Happiness*, the most thorough historical study of philosophical theories of happiness; and the British biographer, memoirist, and historian of art and literature Peter Quennell's *The Pursuit of Happiness*, recounting aesthetic and creative experiences that show how what he calls the gift of exhilarated seeing contributes to our understanding of happiness.[7] Taken together, these books, written before the empirical studies of happiness in recent decades, illuminate much that has been felt and thought and written about happiness in the past. In turn, they help us think about the abundant research findings now coming to the fore – findings that, I believe, the three authors would have found of absorbing interest. Each of the three books, moreover, conveys the author's personal stake in taking on a topic of such scope, along with a measure of hesitation. Should they proceed historically? Or consider particular questions such as that of the relationship between virtue and happiness? Or perhaps focus on individual authors or artists? It would not be possible to restrict their inquiry in such ways. Instead, unusually alert to the perennial controversies over happiness, they cast their nets widely, shedding new light, in the process, on well-known thinkers and bringing out of obscurity works rarely or never discussed in the context of happiness.

This book is the result of my having set out, years ago, to explore what we can learn about happiness. The more I have had a chance to study the clashing views on the subject and the passionate advocacy it can inspire, the more intrigued I have become by the voices and the personal experiences of those who have embarked on its study – the sages and poets and social theorists and increasingly, in recent decades, the social scientists, health professionals, and neuroscientists. As I have come across one person after another examining the subject, I have recognized similarities in their attempts to define happiness and to consider its relevance to the fundamental philosophical question of how one should live, no matter how profound their disagreement about the answer. I have listened for signs of awareness of such disagreements, with some individuals blithely ignoring challenges from outside their own field of expertise, at times moving to silence all critics, while others relish dialogues with friends and adversaries, present and past, whether in history, philosophy, religion, the arts, or the sciences – and, in so doing, invite us to strive to reach beyond our own perspectives.

II

EXPERIENCE

How tell what was neither said nor done nor even thought, but only tasted, only felt, without any other object of my happiness than this emotion itself? I rose with the sun and I was happy; I went walking and I was happy; I saw Maman and I was happy; I left her and I was happy; I went through the woods and over the hill-sides, I wandered in the valleys, I read and I loafed; I worked in the garden, I picked fruit, I helped with the housework, and happiness followed me everywhere.

Jean-Jacques Rousseau, Confessions, 1782[1]

I have often been asked, by kind friends to whom I have told my story, how I felt when I first found myself beyond the limits of slavery; and I must say here, as I have often said to them, there is scarcely anything about which I could not give a more satisfactory answer. It was a moment of joyous excitement, which no words can describe. In a letter to a friend, written soon after reaching New York, I said I felt as one might be supposed to feel, on escaping from a den of hungry lions. But, in a moment like that, sensations are too intense and too rapid for words. Anguish and grief, like darkness and rain, may be described, but joy and gladness, like the rainbow of promise, defy alike the pen and the pencil.

Frederick Douglass, My Bondage and My Freedom, 1855[2]

BLISS, JOY, ELATION, CONTENTMENT, PLEASURE, EUPHORIA, HAPPINESS, ecstasy: how people describe their experience of these states of mind is so much more vivid than efforts to define or explain them.

We need little imagination to enter into the state of mind that Rousseau conveys; or to share, almost viscerally, Frederick Douglass's burst of exultation at finding himself beyond the limits of slavery.

We couldn't even begin to respond to such accounts, however, unless we were able to recall experiences of happiness in our own lives. The recall can be near instantaneous. I have found that when people ask what my book is about, and I tell them, the very mention of the word "happiness" almost invariably leads them to hesitate, as if to look inwards, then recount some intense experience of being overcome by happiness – a great burden lifted, maybe, an unexpected piece of good fortune, a blinding insight, or a deepened awareness of great beauty. A friend told of being caught in a blizzard in her car on a highway, freezing bitterly for hours after her battery died, then feeling flooded with bliss and gratitude at being rescued. For another, the memory that at once came to mind was the feeling of amazed tenderness at the first sight of her newborn baby lying beside her. And a conductor mentioned that his deepest joy would come not while conducting a magnificent piece, as by Beethoven or Mahler, but while sitting quietly going over the score and entering into the music most fully.

Reflecting on such accounts adds to our understanding of the intricate register of feelings and experiences that the very word "happiness" can elicit. Long before we encounter definitions of happiness, joy, and related concepts, most of us learn to distinguish between the nuances and the different durations and intensities in our own awareness of such states of mind. The same is true of sadness, sorrow, melancholy, despair, pain, misery, grief, and agony. The sorting out of the complex shadings and differences among these experiences begins early in childhood. It is aided by responses to other living beings, to nature, to beauty, to the life of the imagination through stories and myths. We gradually come to recognize such distinctions, much as people who dwell in Arctic regions learn to differentiate between a multitude of terms for the varieties of snow and ice.

Unlike distinctions so dependent on climate and natural environment, however, those concerning the many forms of happiness and unhappiness are part of our common *human* environment, regardless of where we live. To be sure, heredity, culture, and sheer luck contribute to differences in how people experience such forms and in turn how they are expressed. True, too, a few individuals seem incapable of perceiving such experiences in others, much less of seeing them as shared. But most people have a rudimentary capacity for feelings of comfort, affection, and joy in interaction with others. We can think of ourselves as wired for such experiences, many of which we also share with a number of animal species. Infants the world over express a range of similar responses of satisfaction, security, even bliss from touch, sound, and smell; and parents everywhere use the same sing-song intonations to elicit such responses.

Charles Darwin held that expressions of basic emotions such as those of joy, anger, disgust, and grief were universal; and that joy, when intense, leads everywhere to movements such as dancing about, clapping hands, and laughing out loud.[3] Psychologist Paul Ekman, expanding on Darwin's observations, has corroborated his conclusions by showing a set of photographs of faces expressing happiness, disgust, surprise, fear, sadness, and anger to people in cultures across the world. A cheerful, broadly smiling face was one that a majority in each society could recognize as happy. The same was true of the pictures that showed sadness and those that showed disgust.[4]

When neuroscientists use brain imaging to track changes in blood flow and neuron activity as people report experiencing pleasure and pain, joy and distress, they also find support for another of Darwin's ideas: namely, that human beings react with empathy on perceiving another's happiness, misery, fear, and other emotions. More generally, recent studies of so-called "mirror neuron systems" in primates and some birds have shown these animals responding to the actions of others as if they themselves were acting thus. Similar mirroring systems at work in the human brain link to the empathy people experience when they witness an emotion such as happiness in others.

Even infants are far better able to respond to facial expressions of shadings of happiness or misery than was once believed. According to Jerome Kagan, "Two year old children have a capacity to infer the thoughts and feelings of another and will show signs of tension if another person is hurt, or may offer penance if they caused another's distress."[5] But while most children are endowed with such a capacity, it is absent in some, due to brain or other impairments; and it can atrophy in others, as a result of abuse or of exposure to unendurable violence. Adolescents already exhibit great differences in their sensitivity to nuances of happiness and unhappiness, as well as in the empathy that enables them to respond to these feelings in others. The same variations occur among adults. Some failures of empathy involve entire categories of people: enemies, for example, or members of racial, religious, and ethnic groups; even half the human race, as when Aristotle and Schopenhauer held that women had less capacity for happiness than men.

By what means do we become capable of perceiving even our own experiences of happiness or well-being? Neurologist Antonio Damasio suggests the metaphor of an ongoing "movie in the brain" that constitutes consciousness: a rough metaphor requiring us to "realize that the movie has as many sensory tracks as our nervous system has sensory portals – sight, sound, taste and olfaction, touch, inner senses, and so on."[6] The experience of happiness can derive from just one or two of these sensory portals or from most of them together. And when people feel unable to experience any form of happiness whatsoever – in so-called states of *anhedonia* – it is as if all such portals were closed to pleasure.

The world's languages convey the different shadings and nuances of words related to experiences of happiness and unhappiness. Every language, of course, has its own set of distinctions, and, as with the translation of any complex words, it is hard to point to an instance of one-to-one correspondence between the terms used in different languages. Most Indo-European words for happiness, such as the English "happiness," the Greek "eutuchia," the German "Glücklichkeit," and the French "bonheur" are based on the concept of "good fortune" which often leads to the feeling of happiness. But it is impossible to draw sharp lines, an authority on these languages points out, "between the pleasurable emotions expressed by pleasure, joy, delight, gladness, happiness, etc."[7] With such cautions in mind, it is nevertheless possible greatly to enhance our ability to perceive the many different forms and nuances of happiness. Just as in trying to understand any other aspect of human experience, say friendship or anger, so with the range of experiences of happiness. Some people respond far more deeply, broadly, intensely to what they experience than others, much as some respond more intensely to what they see than those who are color blind or have weak eyesight. Part of seeking to learn about the experience of happiness should involve asking ourselves how we can come to perceive it more vividly in its many forms. As with vision, there are methods and a variety of aids to deepen and extend our perception of happiness or its absence in our own lives and in those of others, including introspection, drawing on auto-biographies, journals, and other self-narratives, or on works of art, scientific research, and thought-experiments.

Introspection

Introspective observation, as William James underscored in his *Principles of Psychology*, is what we have to rely on *first, foremost, and always* for insight into how we try to grasp the nature of experience. James, who insisted that introspective observation should be coordinated with psychological research, including neurological studies, would have been fascinated to learn of the collaborative work now being done by neuroscientists and psychologists using so-called "neuro-imaging techniques." By means of functional magnetic resonance imaging (fMRI), for example, scientists document the rate of blood flow in different parts of the brain; they can then explore whether images of, or verbal expressions about, particular experiences of pleasure or pain, happiness or unhappiness stimulate specific areas and electrical pathways in the brain.[8] Other researchers have used such techniques to study ways in which damage to certain parts of the brain can result in impaired self-awareness and how this impairment, in turn, diminishes the ability to recognize emotions such as happiness or sadness in other people's faces.[9]

In recent years, neuroscientists, some invoking James, have also studied the introspective practices of meditation of Tibetan Buddhist monks, examining the brain patterns associated with the monks' reports of attaining exceptionally high and lasting levels of happiness. Since 1992, groups of Buddhist monks have visited the Laboratory of Affective Neuroscience at the University of Madison in Wisconsin, to take part in electroencephalographic studies by psychologist Richard Davidson and his colleagues. In such studies, monks who have undergone training in meditation are asked to meditate in the laboratory, their heads wired with EEG electrodes, along with control subjects who have no such training. The brain patterns of monks are consistently found to differ sharply from those of the control subjects, with the monks managing to bring about and maintain levels of activity in the left prefrontal cortex – an area of the brain associated with positive feeling – far above anything achieved by volunteers. The monks, according to Davidson, "are the Olympic athletes, the gold medalists, of meditation."[10]

One of the monks, Matthieu Ricard, who has worked with Davidson throughout, has been hailed by enthusiastic observers as one of the happiest persons on earth. Owen Flanagan, a philosopher in the forefront of efforts to bridge the gap between Western and Eastern philosophy and social and natural scientists, goes so far as to write that if asked whether Ricard was the happiest person ever to exist, he would say "Yes."[11] Ricard himself was trained as a biologist at the Institut Pasteur in Paris before leaving, at twenty-six, for Darjeeling in India to study with a Tibetan master. He has traveled with the Dalai Lama, serving as his interpreter, and reached large audiences through speaking and writing on Tibetan Buddhism and the search for happiness. In his book *Happiness*, Ricard contrasts his earlier life to the sense of flourishing he has come to experience at every moment of his life – a change that he attributes to his good fortune in meeting remarkable people who were both good and compassionate.[12]

William James did not have today's sophisticated methods of examining the brain. But few students of happiness have drawn more meticulously yet imaginatively than he on introspection, through observation and experimental findings as well as on autobiographical writings and works of art. Exploring "the stream of experience," he saw happiness as its central focus. In *The Varieties of Religious Experience* (1902), James illuminated the various ways in which people focused on happiness and unhappiness through meditation, prayer, and soul-searching in different spiritual traditions. He described how, in some doctrines, mere thoughts of happiness, especially if related to sexuality, were to be shunned as grave sins; and explored the different forms that saintliness could assume, whether related to happiness on earth or after death. How to gain, how to keep, how to recover happiness, is for most people, at all times, he held, "the secret motive of all they do, and of all they are willing to endure."[13] As for himself, he had few religious certainties, though he added, in a Postscript, that he agreed in principle with the Buddhist doctrine of karma.[14]

The problem James faced, he wrote to a friend, was how he might best defend "experience" against "philosophy" as being the real backbone of the world's religious life: all that was immediately and privately felt against the philosophers' high and abstract views of our destiny and the world's meaning.[15] To do so, he drew on his vast library of "human documents" or personal accounts of every sort of experience. He presented excerpts from the self-narratives of Teresa of Avila, Rousseau, Thoreau, Tolstoy, and a wealth of others, exploring them from psychological, religious, medical and philosophical perspectives. A great many more such documents have come to light since his time, invaluable for the study of happiness.

Self-narratives

[Prince Amu-an-shi] placed me before his children, he married his eldest daughter to me, and gave me the choice of all his land, even among the best of that which he had on the border of the next land. It is a goodly land: Iaa is its name. There are figs and grapes; there is wine commoner than water; abundant is the honey, many are its olives; and all the fruits are upon its trees; there are barley and wheat, and cattle of kinds without end. This was truly a great thing that he granted me, when the prince came to invest me, and establish me as a prince of a tribe in the best of his land.... I passed many years, the children that I had became great, each ruling his tribe. When a messenger went or came to the palace he turned aside from the way to come to me; for I helped every man. I gave water to him who was thirsty, I set on his way him who went astray, and I rescued the robbed.

Prince Sinuhit, Egyptian papyrus, c. 2000 BC[16]

I stood for a moment in complete silence broken only by the note of a single bird and the susurration of the breeze in the wayside grasses. It was one of those moments of happiness and contentment which give reality to death, since however long we live, there are never enough springs.

P. D. James, Time to Be in Earnest, 2000[17]

In every culture, individuals have spoken or written of their own experiences of happiness and unhappiness, not only in the great classics of self-portraiture, but also in countless memoirs, journals, autobiographies, and confessions. Seen together, these works offer vivid demonstrations of the range and depth and richness of such experiences, invaluable for anyone seeking to go beyond purely personal memories and efforts at introspection.

These self-narratives have flourished well beyond the West and not only in modern times. Lionel Trilling offered an idiosyncratic explanation for the efflorescence of autobiographical writing in late sixteenth- and early seventeenth-century Europe – that "something like a mutation" took place then and that "at a certain point in history men became individuals."[18] Trilling claimed that these men supposed for the first time that they might be of interest to others not for their achievements but simply as individuals; but he gives no evidence for believing that earlier autobiographical writers such as Plato, St Augustine, Seneca, or Marcus Aurelius, along with many less well-known Greeks and Romans had not viewed themselves thus.[19] Nor could he have had in mind the focus on the individual self by twelfth-century anchorites or in fourteenth-century works such as those of Julian of Norwich or her contemporary in Japan, Lady Nijo.[20] And why should anyone imagine that Prince Sinuhit, proudly telling of his marriage and his fertile lands over four thousand years ago, somehow did not think of himself as an individual?

Edith Hamilton, comparing the ancient Egyptians to the Greeks, drew an equally arbitrary line, this time between entire cultures she took to be either capable or incapable of joy. While "Egypt submitted and suffered and turned her face toward death," Hamilton claims, the Greeks

resisted and rejoiced and turned full-face to life. For somewhere among those steep stone mountains, in little sheltered valleys where the green hills were ramparts to defend and men could have security for peace and happy living, something quite new came into the world; the joy of life found expression. Perhaps it was born there, among the shepherds pasturing their flocks where the wild flowers made a glory on the hillside; among the sailors on a sapphire sea washing enchanted islands purple in a luminous air. At any rate it has left no trace anywhere else in the world of antiquity.[21]

So much, apparently, for expressions of the joy of life not only among Egyptians but also among the inhabitants of the Roman Empire, let alone the rest of the ancient world. Fortunately, we now have abundant access to self-narratives as well as extensive documentation from anthropologists and psychologists to rebut generalizations about cultures in which joy is and is not to be found – confirming, instead, Darwin's views on the ubiquity of such emotions.

Even as we set aside limiting notions about self-narratives and individuality and the capacity for happiness, however, we have reason for healthy skepticism about the veracity of claims people make when writing about themselves. Authors, mindful of such skepticism, often begin by insisting that what they write about their experiences will be completely truthful and remarkably open. Thus Montaigne declares, in introducing his *Essays*, that if only custom allowed, he would gladly have portrayed himself "entire and wholly naked"; and Queen Christina of Sweden, in her exceptionally insincere *Autobiography*, asks God to ensure that all she says will bear witness to the truth, however damaging to herself.[22] The critic Philippe Lejeune has pointed to a paradox: on the one hand, the more autobiographers insist on their veracity, the more readers look for discrepancies between the written life and what they know of the author's life. But when, on the other hand, they read works of autobiographical fiction, they try their hardest to discern similarities between the author and the central character.[23]

It is when we consider autobiographical writings from different societies and different periods *together* that they contribute most to our understanding of the different forms that the experience of happiness and unhappiness can take. Looking at all manner of personal accounts together, from letters and journals to memoirs and confessions, can illuminate particular experiences of happiness. Consider the variety of overlapping ways in which sensuous, aesthetic, and spiritual bliss combine for Teresa of Avila:

> In all these types of prayer which I have described in speaking of this last-mentioned kind of water, which comes from a spring, the glory and the repose of the soul are so great that the body shares in the soul's joy and delight. . . . In the whole of the prayer already described, and in each of the stages, the gardener is responsible for part of the labour; although in these later stages the labour is accompanied by such bliss and consolation that the soul's desire would be never to abandon it: the labour is felt to be, not labour at all, but bliss.[24]

And compare the very different pantheistic ecstasy that Thoreau describes in his journal:

My life was ecstasy. In youth, before I lost any of my senses, I can remember that I was all alive, and inhabited my body with inexpressible satisfaction; both its weariness and its refreshment were sweet to me. This earth was the most glorious musical instrument, and I was audience to its strains. To have such sweet impressions made upon us, such ecstasies begotten of the breezes! I can remember how I was astonished.[25]

In turn, the timeless ecstasy that Vladimir Nabokov evokes in *Speak, Memory* differs sharply from that conveyed by Teresa and, in more subtle ways, more focused on aesthetic experience, from that described by Thoreau. "I confess I do not believe in time", Nabokov writes:

I like to fold my magic carpet, after use, in such a way as to superimpose one part of the pattern upon another. Let visitors trip. And the highest enjoyment of timelessness – in a landscape selected at random – is when I stand among rare butterflies and their food plants. This is ecstasy, and behind the ecstasy is something else, which is hard to explain. It is like a momentary vacuum into which rushes all that I love. A sense of oneness with sun and stone. A thrill of gratitude to whom it may concern – to the contrapuntal genius of human fate or to tender ghosts humoring a lucky mortal.[26]

The experiences that Teresa, Thoreau, and Nabokov describe differ from the joyous excitement at achieving freedom in the passage by Frederick Douglass, quoted above, or in those of Archbishop Tutu and Ingrid Betancourt, cited earlier. These states, in turn, are utterly different from the thrill of seeing the possibility of intellectual or artistic achievement, described by Charles Darwin in his *Autobiography*:

The geology of St Jago is very striking yet simple: a stream of lava formerly flowed over the bed of the sea, formed of triturated recent shells and corals, which it has baked into a hard white rock. . . . It then first dawned on me that I might perhaps write a book

on the geology of the various countries visited, and this made me
thrill with delight. That was a memorable hour to me, and how
distinctly I can call to mind the low cliff of lava beneath which I
rested, with the sun glaring hot, a few strange desert plants
growing near, and with living corals in the tidal pool at my feet.[27]

Still another distinction in personal accounts is between the
happiness of a particular moment, such as that described by Darwin,
and that felt for longer periods. Today's researchers routinely draw
this temporal distinction, by asking individuals about their current
experiences or those of the day before or, instead, how satisfied they
are with their lives overall. Psychologists have found that we are
more likely, in looking back at our lives, to remember high points
and dramatic shifts in far greater proportion than ongoing stretches
of happiness or misery.

Similarly, most people, as they construct narratives out of their
lives, stress such shifts disproportionately, leaving out of the
account longer stretches of experience as well as less prominent
ups and downs that don't seem to fit with the larger story. By
contrast, individuals who keep journals often try to record their
immediate experiences more accurately. The journal of an
eighteen-year-old French girl, Claire Pic, who wrote in the 1860s,
is a case in point. She alternates instant snapshots of herself with
evaluations of her entire life. Addressing the pages of her journal
that she felt "listened to her patiently, forgot nothing, and were all
hers," she portrays herself in meticulous, often critical, detail,
while weighing possible futures for herself, including marriage
and religious celibacy. A nun serving as one of her spiritual direc-
tors had objected to the very fact of her keeping a diary, thinking
it wrong of her to be thus preoccupied with herself. Claire, writing
in secret about her life, her thoughts, her impressions, sees her
journal as an act of liberation, following in the steps of Montaigne:

Sometimes I experience an exquisite joy in savoring the blessing of
being, not the banal and material existence of eating, drinking,
sleeping, seeing pretty things, hearing sweet sounds, but the

different, delicate happiness of being a distinct part of the great whole, of being oneself a whole with one's own life, one's own impressions, one's own thoughts. It is a beautiful and grand thing, the right God has given us to say "me" and it is an even greater dignity to be capable of thinking.[28]

Philippe Lejeune, long a student of autobiographical works, uses the format of a personal journal for the first part of the book in which he published excerpts from Claire's journal and those of other young girls of the period. He describes being dazzled by four handwritten pages of hers that he received in the mail, and learning much more while spending a year of exploration in "the land of young girls' journals."[29]

However valuable journals and other autobiographical writings may be for illuminating the range and depth and richness of experiences of happiness and unhappiness, it is to works of art that we must turn to see them more fully. Iris Murdoch, the novelist and philosopher (and herself a lifelong journal-keeper) points to the importance of learning to pay attention to deep areas of sensibility and creative imagination, removal from one state of mind to another, shifts of attachments, and love and respect for the contingent details of the world in order to deepen and enrich our understanding of what people say about their lives.[30] But art, she holds, can carry that process still further, enlarging people's sensibility, allowing them to transcend their limitations, and exhibiting to them "the connection, in *human* beings, of clear realistic vision with compassion":

Any story which we tell about ourselves consoles us since it imposes pattern upon something which might otherwise seem intolerably chancy and incomplete. However, human life is chancy and incomplete. It is the role of tragedy, but also of comedy and of painting, to show us suffering without a thrill and death without a consolation.[31]

The critic Robert Fulford refers to William James's claim that the human mind is "a theatre of simultaneous possibilities," in stressing the enrichment that the arts bring to our understanding of human lives: "The arts build the set for that interior theatre and fill the stage with vivid, memorable characters who mingle in memory with people of our own lives. Even if we are otherwise lonely, we go through life in the company of this ever-expanding society of artists, characters, and images, each of them chosen by us."[32] Fulford mentions the characters of Shakespeare, the portraits of Velázquez, the poetry of W. H. Auden, and the music of Duke Ellington as among those keeping him company. We can all constitute our own society of artists, characters, and images; then carry on dialogues with them, experiencing so much more through what they convey than would be possible without them.

This is not to say, as some claim, that literature necessarily enriches people's perception of human experience, whether of happiness or any other. There is just as much variation in how capable individuals are of responding to works of art as to experiences in their personal lives; indeed, the variation is greater still, depending on the artistic medium in question. To use Fulford's examples, some may find that Shakespeare has nothing to say to them or Velázquez to show them, just as they may be utterly tone deaf to the poetry of Auden or to the music of Ellington. And even those who respond most fully to such works of art may draw still more from repeated readings, viewings, listenings.

Just as introspection and autobiographical writings and works of art can help us extend and deepen our understanding of happiness, so can the thought-experiments conjured up by scientists and philosophers. Such experiments offer scenarios stripped down to test specific hypotheses, generating challenges to what is otherwise taken for granted. Robert Nozick's example of the Experience Machine addresses the common belief that experienced happiness is all that should count in life. His example invites us to look inwards, put ourselves in the place of someone asked to choose whether or not to opt for the lifetime of happiness the scenario promises.

Robert Nozick's Experience Machine

Suppose there were an experience machine that could give you any experience you desired. Super-duper neuropsychologists could stimulate your brain so that you would think and feel you were writing a great novel, or making a friend, or reading an interesting book. All the time you would be floating in a tank, with electrodes attached to your brain. Should you plug into this machine for life, preprogramming your life's experiences?

Robert Nozick, *Anarchy, State, and Utopia*, 1974[33]

The Experience Machine has been used as a thought-experiment in countless classrooms the world over to test the question of whether there is more to life, as Nozick put it, than feeling happy. Could we accept the offer, he asks, to be connected to a machine in which we could feel continuously happy, not only enjoying highly pleasurable sensations but also being profoundly satisfied with our lives more generally? We would be able to live in complete, though deluded, happiness, choosing, perhaps, to believe we are making extraordinary contributions to humanity, having a blissful love life, possessing vibrant beauty, goodness, health, or whatever else would give us lasting happiness. Only none of it would be true.

The imaginary nature of the thought-experiment allows us to strip away practical questions about what floating in the tank would do to our bodies over time, how we would find nourishment, or how we could guard against risks that the machine might malfunction or against mistakes by the technicians stimulating our brains. Thus simplified and abstracted, it can test our willingness to choose to live in a blissful state disconnected from reality as ordinarily understood.

Nozick's answer to his question about plugging into the machine for life is a resounding "No." Far more matters to us as human beings, he argues, than what we experience, no matter how pleasant. First of all, we want to *do* certain things, not just believe we are doing them. Second, "we want to *be* a certain way, to be a certain sort of person. Someone floating in a tank is an indeterminate blob. . . . Plugging into the machine is a kind of suicide." Finally, it

> limits us to a manmade reality, to a world no deeper or more important than that which people can construct. There is no *actual* contact with any deeper reality, though the experience of it can be simulated. Many persons desire to leave themselves open to such contact and to a plumbing of deeper significance. This clarifies the intensity of the conflict over psychoactive drugs, which some view as mere local experience machines and others view as avenues to a deeper reality.[34]

When the example is presented to students, however, it turns out that some few, perhaps around 5 percent, do say they are willing to take a chance on the machine.[35] They refuse to think of this choice as a kind of suicide. They are not disturbed by the fact that they will not be doing the things they believe they are doing while in the tank or by the thought that they would no longer be the person they had once been. The same dissent comes from outside the classroom. Ask people serving lifetime prison sentences whether or not they would opt for such an exit from "reality," or individuals debilitated by excruciating pain; at least some would surely accept such deliverance, were it available.*

Nozick took up his thought-experiment repeatedly in later books, conducting, in this way, a dialogue not only with himself from one book to the next but also with those who had taken issue with his conclusions. His eloquent chapter on "Happiness" in *The Examined Life* focuses more specifically on how his example challenges simplistic views about happiness.[36] The question raised by the thought-experiment, he now specifies, is not whether plugging

---

* The film *The Matrix* presents a similar contrast, between living on in "reality" or reentering the Matrix, the illusory existence in which people do not know that their lives have been taken over and are controlled by men serving the forces of evil. Cypher, part of a small band of heroic men and women who are still at large, opts for going back into what he knows is unreality rather than continue the group's desperate fight with the guardians of the Matrix.

into the machine "is preferable to extremely dire alternatives – lives of torture, for example – but whether plugging in would constitute the very best life, or tie for being best, because all that matters about a life is how it feels from the inside."[37]

"One of the distressing things about the experience machine, as described," Nozick adds, "is that you are alone in your particular illusion." What use is it to feel happy about oneself, one's artistic genius, one's love life or one's service to humanity if it is all imaginary? For those under the spell of such an illusion his question would make no sense; knowing nothing about the imaginary nature of their experience, they would in no way feel alone. But for Nozick, it would be determinative. It demonstrates that happy experiences and a happy disposition, though undeniably wonderful, are not all that we want our lives to consist of:

> We want to experience other feelings too, ones with valuable aspects that happiness does not possess as strongly. And even the very feelings of happiness may want to direct themselves into other activities, such as helping others or artistic work, which then involve the predominance of different feelings. . . . What we want, in short, is a self that happiness is a fitting response to – and then to give it that response.[38]

In asking us, by means of the thought-experiment, to reflect on what constitutes the best life we could wish to have, Nozick's example, in effect, brings up many of the questions that have been raised about the nature of happiness and about what contributes to it or detracts from it. Should we care whether the experience making us happy is "real" and, if so, "real" in what sense? What role should friendship, love, and altruism have for happiness to be most valuable? Is there more to happiness than feeling happy? And more to life than happiness?

To stretch our subjective perspective on deciding whether to enter such a machine, Nozick suggests that we step back to ask how we would answer if our own children's future were at stake: "We would not wish for our children a life of great satisfactions that all depended upon deceptions they would never detect."[39] Part of the power of such a shift away from our strictly subjective perspective is that even persons who may think their own lives so miserable as to be willing to settle for purely imaginary bliss may nevertheless aspire to more for their children. True, parents can have coercive and exploitative plans for their children. But many want their children to be free to mature, to make choices in their lives, to seek meaning, and to experience connectedness to others, including themselves. The Experience Machine eliminates all choice, all concern for a meaningful life, all reciprocity, as well as the capacity to exercise genuine empathy.

Perspectives

This book was written in good faith, reader. It warns you from the outset that I have set myself no goal but a domestic and private one. . . . I want to be seen, here, in my simple, natural, ordinary fashion, without straining or artifice; for it is myself that I portray.

Michel de Montaigne, "To the Reader"[40]

A second way in which looking at autobiographical and fictional accounts and envisaging thought-experiments can help refine and deepen our understanding of experiences of happiness and unhappiness concerns the differences of perspective that they convey. In his *Essays*, Montaigne addresses both himself and readers. This allows readers to experience the dual perspective shaping all personal accounts: that of the speaker or author and that of the listener or reader.[41]

Attending so closely to his own experience, Montaigne stresses all that might cause one to misread oneself and others, misinterpret events and emotions, misremember one's past experience and misjudge future prospects. Likewise, in reflecting on Socrates, Aristotle, Epicurus, Seneca and many other thinkers, he pays as much attention to what he knows of their experiences as to their views. Throughout, he maintains a questioning, skeptical stance, aware at all times of the human propensity to error, bias, and misinterpretation. Testimony from all sources must be given a hearing, he insists; never simply taken on faith. "Even if we could be learned with other people's learning, at least wise we cannot be except by our own wisdom."[42]

Much to the dismay of Church authorities, Montaigne pursues his inquiry into human experience by spreading out his views as in a country marketplace. Sexuality, death, overeating, public service, solitude, pain from gout, hypocrisy, friendship – there is little that he does not bring in, look over from every perspective; turn inside out at times: "I present myself standing and lying down, front and rear, on the right and on the left, and in all my natural postures."[43] In the process, he describes a multitude of experiences of felicity, contentment, discomfort, and acute misery, even the bliss of sudden release from great pain. Is there anything so sweet, he asks,

as that sudden change, when from extreme pain, by the voiding of my stone, I come to recover as if by lightning the beautiful light of health, so free and so full, as happens in our sudden and sharpest attacks of colic? Is there anything in this pain we suffer that can be said to counterbalance the pleasure of such sudden improvement? How much more beautiful health seems to me after the illness, when they are so near and contiguous that I can recognize them in each other's presence in their proudest array, when they vie with each other, as if to oppose each other squarely![44]

Although Montaigne could survey the luxuriant proliferation of quests for happiness and wisdom all around him with a generous tolerance for every form of diversity, he drew the line at persecution, torture, and other violations of fundamental moral values, no matter what the presumed benefits to be derived. He quotes the Roman historian Livy: "Nothing is more deceptive in appearance than a depraved religion, in which the will of the gods is offered as a pretext for crimes."[45] We must see such crimes for what they are, Montaigne advises, while leading our own lives most fully and humanely. He could question every dogma, every certainty, without losing his bearings precisely because he held to the common moral core of fundamental values that allowed him to retain his self-respect. There never was any "opinion so disordered," he had insisted in an earlier essay, "as to excuse treachery, disloyalty, tyranny and cruelty, which are our ordinary vices."[46]

Montaigne aims to partake of the sources of happiness all around him, but without blocking out any form of suffering or the human folly and cruelty that bring so much of it about. As he puts it in his last essay, "Of Experience,"

> When I dance, I dance; when I sleep, I sleep; yes, and when I walk alone in a beautiful orchard, if my thoughts have been dwelling on extraneous incidents for some part of the time, for some other part I bring them back to the walk, to the orchard, to the sweetness of this solitude, and to me.[47]

"We are great fools," Montaigne goes on to say. Why criticize someone who spends his life in idleness or who regrets having done nothing all day? "What, have you not lived? That is not only the fundamental but the most illustrious of your occupations. . . . Our great and glorious masterpiece is to live appropriately."[48] To live appropriately takes resilience, fortitude, curiosity, empathy, but also sheer joy, even in the face of cruelty, suffering, and death. Stanley Cavell explains, in *Pursuits of Happiness*, that he takes "Montaigne's alternative to horror (at human cruelty) to be the achievement of what he calls at the end a gay and sociable wisdom."[49] This achievement was always a work in progress, demanding that he take full account of all the suffering he witnessed at close hand even as he devoted the closest attention to the beauty and joy of so many aspects of daily life.

Ever since Montaigne published his *Essays*, readers have carried on their own imagined dialogues with him.[50] The reason, Virginia Woolf suggests, is that Montaigne managed to give the whole map, weight, color, and circumference of the soul in its variety, its imperfection. "By means of perpetual experiment and observation of the subtlest he achieved at last a miraculous adjustment of all these wayward parts that constitute the human soul.

He laid hold of the beauty of the world with all his fingers. He achieved happiness."[51]

Entering into dialogue in such ways with individuals who convey their personal experience of happiness can be invaluable. Staying only with one's subjective perspective on happiness as on anything else shackles understanding; but so does leaving that personal perspective aside in the name of objectivity. Efforts at complete objectivity that ignore subjective influences invite as much misunderstanding and misinterpretation as does remaining within a strictly subjective point of view. It matters to learn to shift between the two, considering experience as seen both from within and from without.

The attempt to do so inheres in the most fundamental shift in moral perspectives: that of trying to put oneself in the place of those affected by one's actions, so as to counter the natural tendency to moral myopia. The Golden Rule, formulated the world over, takes such a shift to be necessary to the exercise of even the most rudimentary moral choice: to "Do unto others as you would have them do unto you."[52] A related shift in perspectives is that between oneself as a single individual and as one among others. Under what circumstances is it legitimate for me to consider only my own happiness, or only that of those close to me? And when does such a focus do injustice to others? When does it violate their rights?

Still other shifts of perspective involve the direction of time: looking back at all that has come together to shape a situation, then forward to all that may follow it; or surveying an entire lifetime, then focusing on a particular moment or period. Some autobiographers look back at their lives as if they were already dead. David Hume, in his brief "My Own Life", after revealing that he was dying of a bowel disorder, held that the past tense was therefore "the style I must now employ in speaking about myself, which emboldens me the more to speak about my sentiments." Confronting death had brought no abatement in his spirits. Indeed, if he had to point to a period in his life that he would most choose to come back to once more, it would be this last one: "I possess the same ardour as ever in study, and the same gaiety in company."[53]

From our earliest years, most of us experience these different perspectives in at least rudimentary form. Though we rarely reflect upon them, we learn to shift between them intuitively. We know what is meant when people are called happy and when stories end with the central characters living "happily ever after"; but we also empathize with the inward intensity of certain ephemeral experiences of happiness, as in Ralph Waldo Emerson's singling out another such experience in space and time: "Crossing a bare common, in snow puddles, at twilight, under a clouded sky, without having in my thoughts any occurrence of special good fortune, I have enjoyed a perfect exhilaration. I am glad to the brink of fear!"[54]

The same is true of this passage from Virginia Woolf's "A Sketch of the Past":

> If life has a base that it stands upon, if it is a bowl that one fills and fills and fills – then my bowl without a doubt stands upon this memory. It is of lying asleep, half awake, in a bed in the nursery at St Ives. It is of hearing waves breaking, one, two, one, two, and sending a splash of water over the beach; and then breaking, one, two, one, two, behind a yellow blind. It is of hearing the blind draw its little acorn across the floor as the wind blew the blind out. It is of lying and hearing this splash and seeing this light, and feeling, it is almost impossible that I should be here; of feeling the purest ecstasy I could conceive.[55]

As Woolf reflects, in *Moments of Being*, on her efforts to visualize childhood scenes, she tries to pinpoint experiences that neuroscientists and psychologists are still struggling to elucidate:

> In certain favourable moods, memories – what one has forgotten – come to the top. Now if this is so, is it not possible – I often wonder – that things we have felt with great intensity have an existence independent of our minds; are in fact still in existence? And if so, will it not be possible, in time, that some device will be invented by which we can tap them? I see it – the past – as an avenue lying behind; a long ribbon of scenes, emotions. There at

the end of the avenue still, are the garden and the nursery. Instead of remembering here a scene and there a sound, I shall fit a plug into the wall; and listen in to the past.[56]

Through shifts in perspective, we can learn to move between the past and the present, the particular and the general, the tree and the forest, the microcosm and the macrocosm, the case history and the practice of which it is a part – trying to see both surface and depth while losing track of neither. Sometimes efforts to look at autobiographical and other personal accounts from such different perspectives resemble what we do when we walk around a statue to see it from different angles, trying to find the best point of view from which to examine it. Thinking, perhaps, of Montaigne's claim to present himself "standing and lying down, front and rear, on the right and on the left, and in all my natural postures," La Rochefoucauld suggested that "Men and their affairs have their proper point of perspective. Some can only be truly judged from nearby, while fully to appreciate others, distance is essential."[57]

Ignoring such differences in perspective can skew efforts to understand the manifold experiences of happiness just as much as conflating the many kinds of sensory awareness of happiness. We evaluate the empirical findings on happiness by comparing them with what we have learned for ourselves and from what others convey in autobiography and literature and art, and by envisaging thought-experiments. Reflecting on our own experiences and on those of others does not disprove what philosophers, theologians, and social scientists have found, but it does help us appreciate what to question, where to ask for further proof, when to look especially carefully at the methodology and reasoning of the scholars and researchers.

Without taking into account self-narratives and works of art such as those discussed in this chapter, it is easy to block out the differences in felt happiness and perspective that their authors go to such pains to describe; and thus to slip into unreflective, one-dimensional conclusions about the extent to which marriage, for example, or religious belief or health, correlates with happiness: yet another type of what the literary critic I.A. Richards called "premature ultimates" that bring investigation to an end too suddenly.[58]

Short-circuiting besets not only efforts to understand the experience of happiness but also debates about how best to define it and about what contributes to it. Learning to appreciate the subtle differences in feeling and experience that lie behind the many generalizations about happiness makes it necessary at least to pause to ask why all forms of happiness should be lumped together in any one definition or general claim. Few definitions correspond to our everyday experience of happiness, and some are incompatible with one another; yet taken together they, too, require our attention as we seek to encompass the full scope of happiness and to fathom its meaning.

III

DISCORDANT DEFINITIONS

Defining Happiness: A Futile Effort?

Oh, Happiness! our being's end and aim!
Good, Pleasure, Ease, Content! Whate'er thy name:
That something still which prompts th'eternal sigh,
For which we bear to live, or dare to die,
Which still so near us, yet beyond us lies,
O'er-look'd, seen double, by the fool, and wise.

Alexander Pope, *An Essay on Man*, 1733[1]

IN WRITING ABOUT HAPPINESS AND THE ROLE IT PLAYS IN HUMAN lives, Alexander Pope surveyed texts by philosophers, poets, theologians, and scientists from earliest times. His long poem on the human condition was widely read and debated throughout the eighteenth century.* Pope made a crucial distinction: we see happiness both "so near us, yet beyond us." We know the experience intimately; yet the effort to agree on how to encompass it,

---

* Voltaire admired the poem greatly, responding with his own "Discours en vers sur l'homme"; and Kant referred to it appreciatively in a number of texts. Others disparaged it as cavalierly as Pope had written off the views of past philosophers. In his "Life of Alexander Pope," Samuel Johnson was scathing about what he saw as the banality of Pope's claims, no matter the "dazzling splendour of imagery" of his poem. Johnson did not take up any of Pope's arguments in his essay, simply describing them as put forth with "seductive powers of eloquence." One has to look to Johnson's poem, *The Vanity of Human Wishes*, for his own somber views about human happiness. His *Dictionary* merely defines "happiness" as "felicity" and vice versa – hardly an improvement on Pope's "Happiness is Happiness."

fathom its scope, delimit it in words, seems out of our reach. Individuals the world over do have an immediate understanding of happiness – their own and that of others. They have no trouble in distinguishing between photos of happy, sad, and disgusted faces. But definitions of happiness – or for that matter of sadness or disgust – elicit no such unanimity.

Pope summarily disposed of most definitions of happiness in his poem, concluding that past thinkers had simply failed to arrive at a tenable definition:

> Ask of the Learn'd the way, the Learn'd are blind,
> This bids to serve, and that to shun mankind;
> Some place the bliss in action, some in ease,
> Those call it Pleasure, and Contentment these . . .
> Who thus define it, say they more or less
> Than this, that Happiness is Happiness?[2]

By dismissing past thinkers as "blind," Pope conveyed his impatience with the plethora of conflicting definitions he had come across. To him, the matter was far simpler. After sweeping aside pronouncements by "the Learn'd," he declared that happiness has been ordained by Heaven to be available to all in the afterlife, so long as they do what is right on earth. Because order is Heaven's first law, some must be rich and others poor. The poor should be satisfied with their lot; so long as they are virtuous, they will be rewarded in Heaven, where full equality reigns. Only fools imagine that wealth brings greater happiness. Lasting equality and joy after death demand acceptance of inequality and suffering until then. "Whatever is, is right."[3]

To return, then, to the question asked above – Is it futile to seek to define happiness? Absolutely not, I suggest. The effort is futile neither in the sense of "useless" nor in that of "unproductive" as we consider the different, at times clashing, definitions offered over the centuries. Definitions of happiness do express views for which many "bear to live or dare to die," as Pope had written. They are invoked to justify the noblest as well as the most inhumane practices. It makes little sense to turn aside, as he did, from the debates about the nature of happiness that began in antiquity and continue to our day.

In his book *The Idea of Happiness*, V. J. McGill was struck, like Pope, by the conflicting views of happiness from Plato and Aristotle on; but, unlike Pope, he wanted to make the case for each of the principal theories as impartially as he could, so that readers might arrive at their own conclusions. Asking whether philosophers who give such very different definitions of happiness could even be talking about the same thing, McGill concluded that they could, and that the controversies stem from discordant refinements of what begins as a shared meaning. "The root meaning appears to be something like this: A lasting state of affairs in which the most favorable ratio of satisfied desires to desires is realized."[4]

Beginning from such a root meaning, the different conceptions of happiness clash with respect both to the end state of happiness envisaged and to the means required for achieving this end. For those who, like Aristotle, hold expansive views of happiness, the most favorable ratio is to be obtained by increasing the right kind of satisfied desires, whereas the Stoics and others advocate striving to have the fewest possible needs. The Roman Stoic Seneca, for example, taught that limiting one's desires helps to cure one of the fear of death and of all the evils that fate might hold in store and thus makes possible a state of serene happiness.[5] At the opposite end of McGill's ratio, the nineteenth-century French socialist Charles Fourier claimed that "Happiness . . . consists in having many passions and many means of satisfying them." Never averse to cosmological speculation, Fourier added that "We have few passions and hardly sufficient means to satisfy a quarter of them. This is why our globe is for the moment one of the most miserable in the universe."[6]

At every point along this spectrum, philosophers have also differed with respect to the role they grant to pleasure. Whereas most view pleasure as essential to happiness, McGill points to the difference between hedonists such as Epicurus, for whom pleasures alone count for there to be happiness and differ only in quantity, and those who, like Aristotle, hold that pleasures also differ in quality and are far from the only determinants of human action.[7] McGill adopted Aristotle's theory as a framework or template against which to study the different views of how to define happiness – not because he assumed it was true but because he found it to be the most complete theory, the one that asked the most questions, considered the most alternatives, and combined this amplitude with serious attention to consistency and proof.[8]

From the outset of his discussion in the *Nicomachean Ethics*, Aristotle invited dialogue that brought listeners into what he took for granted was an ongoing debate about *eudaimonia* – a term variously translated as human happiness, well-being, thriving, or flourishing.* Early on, he defined it as the highest human good:

> Now we take the human function to be a certain kind of life, and take this life to be the soul's activity and actions that express reason. Hence the excellent man's function is to do this finely and well. Each function is completed well when its completion expresses the proper virtue. Therefore, the human good turns out to be the soul's activity that expresses virtue, and if there is more than one virtue, in accordance with the best and most complete. . . . Moreover, it will be in a complete life.[9]

Sarah Broadie usefully proposes a compact formula for this conception of happiness: "The rational soul's activity of virtue in a complete life."[10] Aristotle included each element in this formula in speaking of happiness as living well and doing well. I shall focus on three questions that have continued to be debated by philosophers and others since his time: Are people the best judges of their own

---

* The Greek term *eudaimonia* derives from *eu*, "good," "well," and *daimon*, meaning "a spirit" or "one's personal fortune," as when Socrates spoke of being guided by his *daimon*. Eudaimonia, therefore, means, literally, "having a good guardian spirit," "a good divine power," or "good fortune." A related term, *eutuchia*, was used more broadly for good fortune, success, or luck; and *makaria* for blessedness, bliss, or the abode of the blessed after death.

happiness, or outsiders? In defining happiness, should we think of entire lives or of shorter periods such as moments, days, or years? And to what extent are virtue and happiness linked?

Are People the Best Judges of Their Own Happiness?

Aristotle took for granted that people's happiness must be evaluated from the outside. Most people identify living well and faring well with happiness, he wrote; but they differ with regard to what happiness is, and the many do not give the same account as the wise. Ever since, people have disagreed about whether people should be sufficient and authoritative judges of their own happiness, or whether it should be determined according to some more objective standard. Examples of definitions of happiness as purely subjective experience abound, such as Willa Cather's statement, inscribed on her gravestone in Jaffrey, New Hampshire: "That is happiness, to be dissolved into something complete and great"; or Thomas Merton's claim that "Happiness consists in finding out precisely what the 'one thing necessary' may be, in our lives, and in gladly relinquishing all the rest. For then, by a divine paradox, we find that everything else is given us together with the one thing we needed"; or Ayn Rand's view that "Happiness is that state of consciousness which proceeds from the achievement of one's values."[11]

Economist Richard Layard, invoking Jeremy Bentham, defines happiness simply as "feeling good – enjoying life and wanting the feeling to be maintained."[12] Many of today's psychologists likewise emphasize the subjective perspective on happiness, either as sensations or as judgments about how one feels about one's life as a whole. Ed Diener contrasts the new studies that ask people how they themselves feel about their level of satisfaction, contentment, and well-being to the traditional focus on objective criteria for happiness: "It is this approach to defining the good life that has come to be called 'subjective well-being' (SWB) and in colloquial terms is sometimes labeled 'happiness.' "[13]

Contemporary philosophers disagree sharply about whether or not happiness should be defined from a purely subjective point of view. Richard Brandt defines happiness in strictly subjective terms, with two components. For people to be happy, he views it as necessary, first of all, that they like those parts of their total life patterns and circumstances that they think are important. Secondly, "One would not call a man happy if he did not frequently feel joy or enthusiasm or enjoy what he was doing or experiencing."[14] And Wayne Sumner proposes that "happiness, or life satisfaction, is a positive cognitive/affective response on the part of a subject to (some or all of) the conditions or circumstances of her life."[15]

To all who, like Brandt, define happiness in terms of subjective experience, Aristotle might reply that a life cannot be seen as a happy, thriving one in the sense of *eudaimonia* unless it expresses "a rational soul's activity of virtue."[16] A wise outsider's judgment is needed to weigh whether an individual's life meets the criteria for rationality, activity, and virtue. Today, few philosophers who argue that happiness, well-being, or welfare can be determined by means of "objective lists" include all of Aristotle's criteria. But like him they hold, as Derek Parfit puts it, that "certain things are good or bad for us even if we would not want to have the good things or avoid the bad things."[17] Martha Nussbaum argues for a broad Aristotelian approach that stresses objectivity while recognizing that new circumstances and new evidence may require revisions in what are held to be Aristotelian virtues and the deliberations they guide.[18]

John Kekes combines subjective and objective perspectives on happiness. He defines it in such a way as to include both how individuals view their own level of happiness and outside judgments that they can be wrong or deluded at times. We can say that people's lives as a whole are happy, Kekes holds, "if and only if two conditions, individually necessary and jointly sufficient are met. First, the agents sincerely believe that enough of their important first-order wants are satisfied. Second, knowledgeable observers have no reason to suppose that latent internal defects or inhospitable contexts falsify the agents' sincere judgments."[19] In speaking of "latent internal defects," Kekes addresses problems that philosophers, novelists, and psychologists have illuminated: the degree to which we are often unable to evaluate our situation in a way that would make sense even from our own subjective perspective, as when we misremember the past, misjudge our present experience, and fail to predict how happy we are likely to feel about future events.

When outsiders evaluate what individuals say about their own happiness, two separate kinds of judgment are at issue. The first asks whether people's evaluations of their own happiness are based on true facts – unlike the illusory bliss of those floating in Robert Nozick's Experience Machine, say, or the delusions of someone proudly believing he is Napoleon, or yet of members of the Jonestown or other suicide cults who go blissfully to their death in hopes of still greater future happiness; or unlike much more common self-evaluations such as those by people who wrongly believe their spouse or their children love them or that they are admired and respected at work. On this score, some philosophers hold that such deluded states simply cannot be termed happy, whereas others insist that if people who are ignorant or deceived about the true circumstances of their lives nevertheless say they are happy, then they are, and that is all there should be to it, even from the perspective of outsiders.[20] Robert Nozick refuses to go that far, suggesting that "We would be reluctant to term someone happy at a particular moment or in life in general if we thought the evaluations upon which his emotions were based were wildly wrong."[21] Even he speaks of reluctance rather than refusal, however, given the fierce disagreements throughout history about precisely when people's evaluations of their own happiness *are* wildly wrong.

The second, quite different, judgment brings to bear some outside standard thought indispensable for happiness, such as the possession of virtue or of particular religious convictions. Regardless of what people believe about their own happiness and how right or wrong they may be about such facts as their spouse's affection or their performance at work, the question remains: Can they actually be called happy if they fail to meet that standard? Here again, there can be sharp splits between those who see people's subjective evaluation of their level of happiness as determinative and those who insist that outside evaluations can override and thus invalidate personal views.

John Rawls, while not viewing the possession of virtue or of religious belief as required for happiness, nevertheless includes one outside standard – that of possessing a rational life-plan – as part of his very definition of happiness. Like Aristotle and many other philosophers, Rawls takes for granted that happiness requires such a plan.[22] Citing Josiah Royce's even more restrictive claim that a *person* may be defined as a human life lived according to a plan, he adds that a rational plan of life establishes the basic point of view from which all judgments of value relating to an individual person are to be made and finally rendered consistent. We can think of a person as being happy, Rawls suggests,

> when he is in the way of a successful execution – more or less – of a rational plan of life drawn up under (more or less) favorable conditions, and he is reasonably confident that his plan can be carried through. Someone is happy when his plans are going well, his most important aspirations being fulfilled, and he feels sure that his good fortune will endure. [23]

Having such a plan of life surely contributes to happiness for a number of individuals. Striving for a better plan is worthwhile for most of us. And it can be illuminating to consider individual life stories from the point of view of the degree to which they exhibit such strivings. But as a defining criterion for happiness I find the possession of a rational plan of life needlessly restrictive. Admirable as persons happy in their possession of such plans may be, they are not easy to find, the less so if they are also supposed to feel sure that their good fortune will endure. Indeed, if they do feel this confident, they may be unwisely blinding themselves to the vagaries of fortune. Montaigne, however, may have gone too far in the opposite direction, holding that *no one* could make a definite plan of his life: that we think of our lives only piecemeal, though we can try to keep our plans from going astray by setting up goals for ourselves.[24]

Persons with long time horizons, accustomed to surveying both their past and present lives and to planning for the future, are especially likely to try to make plans for their lives. But why should individuals who, like most adolescents, have shorter time horizons not count as happy so long as they experience their lives as fulfilled and rich in happiness?[25] And what about all those who do have plans or at least hopes for their future, but who are uncertain about whether their plans are going well, about whether their most important aspirations are being fulfilled, or about whether their good fortune will endure? I find the view that happiness calls for our having rational life-plans to be as limiting as the claim that it requires the possession of good health, a good conscience, close friends, or other factors that, as we shall see, have been proposed as indispensable for happiness.

As we seek to understand the scope and depth of human explorations of happiness, it makes little sense to wall off either subjective or objective perspectives. And while I believe that the subjective experience of happiness must have priority, I take both the insider's and the outsider's perspective to be needed for fuller understanding. Each can help counteract errors and biases in the other. Those who take only a person's own experience into account are as prone to such errors and biases as those who focus only on objective indications of people's preferences and needs. Even as we adopt either the subjective or the objective perspective, we should not lose sight of the other.

Happy Moments, Periods, Lives

Then why not say that the happy person is the one who expresses complete virtue in his activities, with an adequate supply of external goods, not just for any time but for a complete life?

Aristotle, *Nicomachean Ethics*[26]

Aristotle had a second argument to buttress his view that a person's happiness is best evaluated from the outside: that we can only determine whether people are happy after their lives are over, given that a great many misfortunes can afflict even the happiest-seeming individuals late in life. "For one swallow does not make a spring, nor does one day; nor, similarly, does one day or a short time make us blessed and happy." Pointing to the calamities that befell old King Priam in the *Iliad*, Aristotle argued that no one could call someone happy who had experienced such events and had ended so wretchedly.[27] But could this mean, he continued, that we could never call people happy during their lifetime, as Solon had reputedly answered King Croesus, who had asked whether anyone was happier than he?[28] And is it even enough to wait until someone has died before deciding whether or not his life was a happy one, given that his children may do poorly or suffer dishonor and in turn bring him dishonor? But how can dishonor damage someone's happiness after death?

Aristotle's questions have echoed in discussions ever since. Few have agreed with him that a complete life is needed for judging a person's happiness, even though many have accepted the weaker claim that neither one day nor a short time can make us fully "blessed and happy." Julia Annas has pointed to child-rearing to support her view that we do think, centrally, of *living* happy lives: "As we bring up our children, what we aim for is not that they have episodes of smiley-face feeling, but that their lives go well as wholes."[29] In so far as what people aim for is long-term happiness, the analogy with planning for one's children is cogent; but many parents may have entirely different, less far-reaching, aims for themselves and for their children.

As for Aristotle's mention of the reversal of fortune for King Priam late in life, few recent examples rival that of businessman Bernard Madoff, who admitted, in November 2008, that he had swindled investors of 50 billion dollars. Older than Priam, richer than Croesus, widely admired for sagacity and philanthropy, Madoff had given every impression, up to then, of leading the happiest of lives. He might well have answered survey questions about subjective well-being in the most positive terms, as would friends and colleagues asked to evaluate his level of happiness. Yet the answer that Solon was said to have given to Croesus applies as devastatingly to Madoff: "I see you are very rich and the king of many men. But I cannot yet answer your question, before I hear that you have ended your life well."[30]

Some believe that it makes no sense to speak about the happiness even of a day, a month, a year, let alone one's life seen as a whole. For example, the Russian dancer Anna Pavlova reputedly used the image of a butterfly rather than that of Aristotle's swallow in holding that happiness is like a butterfly which appears and delights us for one brief moment, but soon flits away. Her contemporary Sigmund Freud, likewise, saw human hopes for lasting happiness as illusory. Happiness in the strictest sense, he held, can only come through brief experiences of satisfied needs.[31] Philosopher and psycho-analyst Jonathan Lear, contrasting the views of Aristotle and Freud, has suggested that Aristotle smuggles in a fantasy of *eudaimonia* or happiness at the beginning of the *Nicomachean Ethics* to seduce us into believing that there can be lasting happiness and that virtue is central to achieving it. People find "real" happiness only in fleeting moments, Lear argues, citing an old English usage of "happiness" in terms of happenstance: "Happiness, on this interpretation, is . . . quite literally a lucky break."[32]

Today's scholars studying subjective well-being see little need to define it either in terms of entire lives or of shorter periods. Most people regard it as obvious that one can be happy for a moment or a day or a year, and that what matters is that one should not use one time period to judge another. Psychologist Daniel Kahneman has distinguished four temporal dimensions for experiences of pleasure or pain: "instant utility," which concerns these at a particular moment; remembered utility; satisfaction with domains of life such as family life or work; and happiness or well-being, encompassing all domains but not necessarily over a lifetime.[33] Focusing on the first two dimensions, neuroscientists examine people's brain responses as they report pleasure and pain and then try to remember the experiences later on.

The notion that there should be one kind of happiness, whether experienced briefly or for longer periods of time, is need-lessly restrictive, considering the variety of human experiences of happiness, of purposes for which people seek it, and of factors thought to contribute to its achievement. We know from experi-ence the momentary happiness that Pavlova and Freud speak of, but can also envisage the reflection on a whole life to which Aristotle refers. International surveys now routinely ask individuals both about short spans of pleasure and longer-term satisfaction with their lives (though not with their entire life span.)[34]

But to all who disagree with Aristotle on the need to consider entire lives, not shorter periods, and on the necessity for wise outsiders' views of people's happiness, he might well reply that they have left out the most important part of his definition of *eudaimonia* as the soul's activity that expresses virtue. And it is true: most of the definitions mentioned above make no mention of virtue or moral worth. It is as if Aristotle were to challenge them with one sharp "Yes but" question on moral grounds. How can you possibly think of human happiness without referring to virtue?

Virtue and Happiness

Does happiness rest on a foundation of virtue, as Thomas Jefferson wrote, quoting Seneca's "On the Happy Life"?[35] For Jefferson as for Plato, Aristotle, and most classical thinkers, the concept of human excellence or virtue (the Greek *arete* and the Latin *virtus*) was inextricably linked to that of happiness. In antiquity, the scope of this concept was broader than moral virtue, extending to human excellence more generally. But opinions clashed regarding just how virtue and happiness are linked. Does happiness rest on a foundation of virtue or is the reverse true? Plato and the Stoics held the first view: if you strive to be virtuous, that is the only way you can attain happiness. Epicurus held the second view: the effort should be to become happy; in the process, you would of necessity have to exercise the virtues.

A stronger version of the first view holds that only the virtuous can be truly happy – that virtue is not only necessary for happiness but sufficient – and, of the second view, that only the happy can be truly virtuous. Many have doubted both stronger versions, needing only to look around to see virtuous people suffering and evil people giving every evidence of being happy. Past thinkers often mentioned Phalaris, the tyrant of Acragas, or Agrigentum, in Sicily, *c.* 570–550 BC. He was the stock figure of evil in antiquity, as Hitler is today. It was said of Phalaris that he had a brazen oven shaped like a bull, in which he roasted his victims, including children, sometimes serving them to unwitting guests as a delicacy. The question presented to those who maintained that a thoroughly wise and virtuous person was by definition happy was: Could such a person remain happy even if forced to enter the brazen oven? Cicero, who held that a thoroughly wise and virtuous individual would indeed be able to step down into the bull of Phalaris, nevertheless refused to agree with Epicurus's improbable claim, that a wise man inside the bull would say "How sweet!"[36] And a separate question was posed to those who, like Epicurus, deemed happiness to be pleasure – *hedone*: What if the tyrant's pleasure is immense? Is it not happiness? If not, why not? How could such a theory rule out the enjoyment, perhaps extreme, felt by torturers and child abusers and serial killers?

Aristotle disagreed with his teacher Plato, who saw virtue as not only necessary but the one factor that was sufficient for happiness. True, Plato had added that virtue also needs, as "instruments for use," good health, strength, and sound senses, along with external advantages such as wealth, good birth, and good reputation. Without these, it will be difficult for most people to exercise virtue in practice, even though "the wise man will be no less happy even when he is without these things."[37] For Aristotle, such benefits were not only useful but utterly indispensable to anyone hoping to lead a virtuous life. He did not think it possible for someone to remain happy, however virtuous, under extreme circumstances such as being tortured on the rack.

Three sorts of goods, Aristotle specified, contribute to happiness: goods of the soul, including moral and intellectual virtues and education; bodily goods, such as strength, good health, beauty, and sound senses; and external goods, such as wealth, friends, good birth, good children, good heredity, good reputation and the like.[38] Some goods, such as money, serve as necessary or especially effective instruments for the doing of virtuous actions, while others, such as not being ugly, put one in a better position for exercising virtuous actions. But the external goods, while necessary, cannot be sufficient for achieving a flourishing life. The merely lucky person "mimics" the virtuous one by sheer luck. Women, slaves, the poor, and the childless cannot attain a state of full flourishing. As for vice, it is sufficient, for Aristotle, to ensure unhappiness – *kakodaimonia* – even if it be ever so abundantly furnished with corporeal and external goods.

With the breaking up of the Greek city-state and growing insecurity for even the most well-endowed individuals, different schools of philosophy stressed the need for independence from goods not under one's own control. Among them, the Greek and Roman Stoics contributed to the longest-lasting tradition of ethical advice for people asking how they might ensure personal happiness in spite of the unreliable and chaotic circumstances in which they might find themselves.[39] Zeno the Stoic warned of the unhappiness bound to befall those who depended on what was not wholly under their control. The only thing completely under one's control was virtue and wisdom combined, permitting full understanding of what to do and how to respond even to the worst calamities. Seneca, Epictetus, and Marcus Aurelius developed and elaborated on this view. Sternest, perhaps, was Epictetus, the former Roman slave, who insisted that the essence of philosophy is to depend as little as possible on external things, and that the only road to happiness is to cease worrying about things that are beyond the power of our will.

Epicurus challenged the preeminence given to virtue by Aristotle as well as the Stoics, by holding that pleasure, *hedone*, was what defined *eudaimonia*, to the point of being identical with it. A happy life is tranquil, simple, loving, and above all free from pain, fear, and suffering, available to all regardless of social status, nationality, or gender. Such a life of pleasure, Epicurus held, would of necessity have to be a virtuous one; injustice, for instance, would be incompatible with it. It would also be inconceivable without friendship: acquiring friends, for Epicurus, was by far the most important thing a wise person could do to ensure happiness throughout the whole of life. True, a wise man should rid himself of needs, including the need for friends, that could render him dependent; but having done that, he could relish their companionship. Such a life could be simple in the extreme. Indeed, most of the "external goods" specified by Aristotle, such as health, good looks, or wealth, mattered little in such a reconfigured view of pleasure as happiness; those who had achieved wisdom and learned to reduce their sensitivity to suffering would be able to be fully happy even while tortured on the rack.

The debates over the ways in which virtue and happiness were linked continued over the centuries. But regardless of how thinkers envisaged that linkage, most took its existence for granted, and agreed that happiness was the supreme goal sought by human beings, the *summum bonum*. It therefore came as a shock when Immanuel Kant declared that the link had to be severed. By arguing that living so as to be *worthy of happiness* was what should matter above all, not happiness in its own right, Kant effectuated a "dethronement of happiness," in McGill's telling phrase, that posed an abrupt challenge to earlier thinkers.

Immanuel Kant's Dethronement of Happiness

In his first major work on ethics, the *Groundwork of the Metaphysics of Morals* (1780), Kant famously disposed of all the various talents, virtues, gifts of fortune, and other factors that philosophers had long discussed as contributing to happiness.* All were worthless, he argued, in the absence of a good will, indeed dangerous when possessed by evildoers:

> It is impossible to think of anything at all in the world, or indeed even beyond it, that could be considered good without limitation except a good will. Understanding, wit, judgment and the like, whatever such *talents* of the mind may be called, or courage, resolution, and perseverance in one's plans, as qualities of *temperament*,

---

* Kant, who knew Alexander Pope's work well and quoted the *Essay on Man* in his *Anthropology* and elsewhere, may have had that poem in mind as well in composing this broadside against conceptions of factors contributing to happiness.

are undoubtedly good and desirable for many purposes; but they can also be extremely evil and harmful if the will to make use of these gifts of nature, and whose distinctive constitution is therefore called *character*, is not good. It is the same with the *gifts of fortune.* Power, riches, honour, even health, and that complete well-being and satisfaction with one's condition called *happiness*, produce boldness and thereby often arrogance, unless a good will is present which corrects the influence of these on the mind . . . so that a good will seems to constitute the indispensable condition even of worthiness to be happy.[40]

True enough, Kant added, qualities such as "moderation in affects and passions, self-control, and calm reflection" make the work of a good will much easier; but however unconditionally these traits were praised by the ancients, they can become extremely evil too, in the absence of a good will. The "coolness of a scoundrel makes him not only far more dangerous but also immediately more abominable in our eyes than we would have taken him to be without it."[41]

Whereas most earlier thinkers upheld the connection between happiness and virtue, Kant's insistence on severing the link between the two was the most forceful set of "Yes but" questions aimed at all such claims. Morality, doing what duty called for, could never be derived simply from the desire for happiness. Instead, the moral law might require individuals to sacrifice all happiness, even life itself. Kant quoted Juvenal on the subject of a soldier, a guardian, or a judge asked to bear witness in a dubious cause: "Though Phalaris himself should dictate that you perjure yourself and bring his bull to move you, count it the greatest of all iniquities to prefer life to honour and to lose, for the sake of living, all that makes life worth living."[42]

Not only was there no such link between morality and happiness as ancient philosophers stipulated; they had also been wrong, Kant held, in thinking that happiness was achievable on earth. To support this view, however, Kant had to set the bar for happiness impossibly high in his very definition of happiness: "a rational being's consciousness of the pleasantness of life uninterruptedly accompanying his whole existence is happiness."[43] Since such uninterrupted lifelong happiness is not achievable in this life, the problem for ethics is that of finding incentives for moral action apart from happiness. Why should people seek to be good, to be worthy of happiness, if happiness itself is out of their reach? Kant did not think there could be proof of an afterlife in which all would be set to rights by a God who would reward the good, punish the wicked. Nevertheless, he believed it was necessary at least to postulate such a future: one in which God might ensure that happiness and virtue would come together, unite, after death. Having severed the link between the two in human life, he hoped they could be re-linked in afterlife. Without postulating such an afterlife, in which "an intelligent author" saw to it that the good would be rewarded however much they had suffered during their lifetime, and the evil persons punished no matter how they had seemed to prosper while still alive, the incentive to act morally would be weakened.

Many thinkers after Kant have, rightly in my opinion, disagreed with his two conclusions: that happiness is not achievable in human life, and that it can therefore be hoped for only in afterlife. Among the critics, Friedrich Hegel's view was that Kant had felt obliged to resort to a *deus ex machina* at the end of the play by conjecturing that happiness would be distributed according to degrees of worthiness in another world. John Rawls suggests that Kant never reworked his philosophical theology so as to make it consistent with his moral philosophy. Matching happiness with virtue cannot be part of the moral law as it applies to us by way of the categorical imperative. James Griffin points out that the answer to the question "Why be moral?" can be that it is either a way or the only way to be happy. It is easier to argue that virtue is sufficient for happiness than that it is necessary; easier to acknowledge, as seems clear from countless examples, that there are people for whom acting morally does suffice to make them happier, than to argue – what many examples contradict – that *only* such people can be happy. And Christine Korsgaard suggests that Kant was wrong in holding that sustaining our moral commitment requires us to postulate an ultimate harmony between nature and morality – in effect the ultimate goodness of the world.[44]

But rejecting one or both of Kant's conclusions is no reason not to take seriously the central thrust of his "dethronement of happiness": his challenge to the common assumptions that people should seek happiness first and foremost and that pursuing it will most often link up with high moral standards, so that happier people are also better, kinder, more honorable; and his assertion, instead, that what matters most is whether or not we deserve to be happy. This involves striving to live up to the highest moral standards rather than merely seeking happiness wherever one can find it. People should regard such standards, he had long insisted, as stricter for themselves than for others. In judging the actions of others, we should always be charitable, mindful of human frailty and infirmity, but never use these facts as "excuses for our own misdeeds."[45]

In the last book Kant published, he blamed philosophers for talking about virtue only in fragments, instead of making it "interesting as a whole to all people in its beautiful form" by singling out the concept of character.[46] For humans who, alone among living beings, have what he called the gift of morality – the extraordinary possibility of making moral choices – the fundamental choice is whether or not to choose to be someone of character out of respect not only for other people but for themselves. In the absence of making that choice, just wishing to become a better person is worthless.

It is not hard to find persons of character, in Kant's sense, who are far from happy, nor some among those he characterized as scoundrels who appear to thrive. Phalaris was thought to have flourished; in our own time, perpetrators of violence and fraud on a colossal scale such as Idi Amin and Bernard Madoff gave every indication of thriving until stopped in their tracks. Bernard Williams suggests, however, that fewer amoral people may be happy than is commonly assumed. Some people are horrible because they are unhappy and sometimes the converse is true. Others are not horrible at all but miserable. Still, there is also the horrible person who is not miserable at all but "by any ethological standard of the bright eye and the shining coat, dangerously flourishing. For those who want to ground the ethical life in psychological health, it is something of a problem that there can be such persons at all. [But] they seem sleeker and finer at a distance."[47]

Williams cautions us against grounding the ethical life too swiftly in psychological health. At the same time, however, he seems prepared to give support to the intuitive notion stressed in moral theories and religious traditions alike, that, on the whole, people will be happier if they also strive to live up to moral standards and unhappier if they do not. In the decades since he wrote, researchers have explored the influence of brain injuries, traumatic experiences, relentless indoctrination (or brainwashing), and sheer habituation on the capacity for empathy and moral judgment. Once individuals are debilitated in such ways, they may be impervious to the suffering they are inflicting; some may even give every appearance not only of psychological health but of exultant happiness.

More often, however, studies have found some correlation between altruism and subjective well-being. People who rank high on scales of subjective well-being are also found to be more likely to engage in certain altruistic activities such as volunteering.[48] Psychologist Stephen Post has concluded that altruistic emotions and behaviors are associated with greater well-being, health, and even longevity.[49] Sonja Lyubomirsky points to scientific experiments which demonstrate that kindness is good for the doer.[50] The experiments are designed, she suggests, to overcome the natural question with respect to the supposed link between kindness and happiness: Are people who engage in helpful activities happier to begin with, or does their felt happiness increase as a result of helping others? She attempts to get around this question by dividing randomly selected subjects into two groups, then asking those in one of the groups to engage in acts of kindness and not those in the other group; the results show at least a temporary increase in subjective well-being for those who do so engage.

Neuroscientists have also found such a temporary increase in a study scanning the brains of volunteers asked to think about a scenario involving either anonymously donating a sum of money to a charitable organization or keeping it for themselves. When participants chose to place the interests of others before their own, this activated reward systems in the brain that usually light up in response to money, food, and sex. Thinking of the interests of others, the researchers concluded, was not so much a moral obligation to be obeyed at some cost to personal satisfaction, but a source of satisfaction in its own right that produced pleasurable responses in the brain.[51]

It is doubtless true that satisfaction with one's own kindness, generosity, and friendliness may contribute to higher levels of subjective well-being. But a closer look at the perceived link between kindness and happiness brings out moral discrepancies, illuminated, above all, in literature. Take the cavalcade of hypocrites and self-satisfied malefactors from the novels of Charles Dickens: the unctuous Mr Pecksniff, in *Martin Chuzzlewit*, for example, a self-proclaimed humanitarian so concerned for others that he named his daughters Mercy and Charity, even as he defrauded everyone in sight while intoning the language of universal love; or Mrs Jellyby in *Bleak House*, so absorbed in distant philanthropic efforts to aid African natives that she alternately neglects and exploits her children and allows her house to sink into disarray.

Even for individuals nowhere near so dastardly or callous, the desire to be kind conflicts in innumerable ways with other moral requirements. It is important, in examining existing research and in planning further research on altruism, to think through its role as only one aspect of moral excellence, sometimes in conflict with other aspects such as honesty and justice. Throughout history, altruism toward members of one's own family, community, faith, or ethnic group has routinely gone along with utter indifference to outsiders and with every kind of brutality inflicted on enemies.

Definitions as Rorschach Tests

"Tell me how you define happiness and I'll tell you who you are!" Such a claim, like John Ruskin's "Tell me what you like and I'll tell you who you are!" and similar ones beginning "Tell me what you eat . . ." and "Tell me who your friends are . . ." has at least a grain of truth. Many definitions of happiness appear to extrapolate from personal experience. Already in the eighteenth century, Denis Diderot noted, after making his way through seemingly endless treatises on the subject, that they are never anything but the history of the authors' own experience of happiness.[52] Definitions of happiness can function, then, as Rorschach tests of sorts – of personality traits, hopes, and biases. Some definitions leap off the page. Nietzsche, for example, took specific issue with philosophical and religious claims about links between happiness and virtue, altruism, religion, and sympathy. He defined happiness as "The feeling that power *increases* – that a resistance is overcome," adding:

> *Not* contentment, but more power; *not* peace at all, but war; *not* virtue, but proficiency (virtue in the Renaissance style, *virtù*, free of moralic acid).
>
> The weak and ill-constituted shall perish: first principle of *our* philanthropy. And one shall help them to do so.
>
> What is more harmful than any vice? – Active sympathy for the ill-constituted and weak – Christianity . . .[53]

Goethe spoke in terms that expressed an utterly different, but equally personal, outlook on life: "The highest happiness of man is to have probed what is knowable and quietly to revere what is unknowable." Ellen Key, the Swedish feminist, educator, and child advocate, drew on Goethe's saying that "happiness is the development of our capabilities," in her 1900 book, *The Century of the Child*; then went on to list memory, the power to think, and "above all feeling" among these capabilities.[54] W. B. Yeats wrote in his diary that "I think all happiness depends on having the energy to assume the mask of some other self; that all joyous or creative life is a rebirth as something not oneself – something created in a moment and perpetually renewed; in a game like that of a child where one loses the infinite pain of self-realization, a grotesque or solemn painted face put on that one may haste from the terrors of judgement."[55]

Among contemporary philosophers, Robert Nozick writes that "We want experiences, fitting ones, of profound connection with others, of deep understanding of natural phenomena, of love, of being profoundly moved by music or tragedy, or doing something new and innovative, experiences very different from the bounce and rosiness of the happy moments."[56] And Stanley Cavell offers a different ideal: "The achievement of human happiness requires not the perennial and fuller satisfaction of all needs as they stand but the examination and transformation of these needs."[57]

Incompatible as many definitions may be, they not only enrich our understanding of the many-layered conceptions of happiness, but also help us reflect on the experiences the authors draw upon, the aspirations they have in mind. Such general definitions, often referring to "all happiness" or to what "we" mean by the term, differ from descriptions in which individuals present their personal views of what makes for happiness – what contributes to their own happiness – without the slightest suggestion of laying claim to universal validity. These, too, can be instructive. Art historian Roy Strong, for example, in his short book *On Happiness* specifies that what gives him happiness may not do so for others:

> For me it is a strange web of things which criss-cross each other: friendship, living each day as though it were one's last, never going back but only forward, looking for the good things in everyone, the passage of the seasons in a landscape or a garden, cooking a good meal and sharing it, being together with the person you love most, trying never to nurse past resentments, filling one's environment with a clutter of things which are the repository of felicitous memory, expecting nothing as a rule of life, so that anything which comes in is an unlooked for pleasure, and above all, as a priest once bade me, to make each day a perpetual awareness of the presence of God.[58]

That there should be so many definitions of happiness is not surprising. As with efforts to define other abstract terms such as "beauty," "love," or "violence," there are no agreed-upon rules for defining them, no established criteria for when the terms do or do not apply in particular cases.[59] Such abstract terms provide ideal vessels into which people can pour quite different, sometimes clashing, meanings. These meanings, unlike the more precise definitions of, say, a triangle or a comet, often constitute *redefinitions* that depart from ordinary understandings of such concepts; and are known as persuasive definitions. The concept of happiness, given the central role it plays in views of life and death and in political and religious doctrines, is especially likely to be redefined for such persuasive purposes.[60]

It is tempting to reject all persuasive definitions as being unscientific or vague, maybe even as "at best a mistake and at worst a lie."[61] But when it comes to happiness, as with beauty or love, those who offer a new definition invite us to think anew about a concept we may have taken for granted. We learn more, in reflecting on happiness, from juxtaposing definitions than from merely taking any one intuitive notion for granted. Some are especially likely to challenge and provoke commonplace views – as with Jonathan Swift's "Happiness . . . is a perpetual possession of being well Deceived";[62] or Nietzsche's "the feeling that power increases," cited above.

As we try to understand the different views of happiness, therefore, it is worthwhile to view them together, including those that seem either strangely limiting or diffuse. And depending on the purposes they are meant to serve, we may find that some definitions, once examined, are not in conflict at all. When, for example, Iris Murdoch has a character in one of her novels say that "Happiness is a matter of one's most ordinary everyday mode of consciousness being busy and lively and unconcerned with self," this need in no way jar with Sonja Lyubomirsky's definition of happiness as referring to "the experience of joy, contentment, or positive well-being, combined with a sense that one's life is good, meaningful, and worthwhile."[63]

Given the multitude and scope of conceptions of happiness that overlap sporadically only to part company, it is important not to exclude any of them summarily, short-circuiting debate about the arguments for accepting one or another. Alexander Pope was wrong in concluding from the fact that "the learn'd" have incompatible views about happiness that no one such teaching can be valid. What is needed, instead, is to look at each one, striving to understand the role it has played in human lives and societies, then weigh the evidence each offers. Combining the effort to understand the different views more fully with caution before assenting to them – above all when they demand radical life changes – is yet another way of balancing empathy and resilience.

As soon as we want to go beyond merely trying to see the depth and scope of the different definitions of happiness to ask about the reasons for their incompatibility, we have to take into account the practical uses to which each might be put. Helping us to understand happiness better, however, is hardly the aim of those thinkers who are most certain that their own definition of happiness is the only legitimate one. There is no one definition of happiness, I suggest, that should exclude all others, much less be imposed by force or indoctrination. We need to look at the different ones together and consider the roles they have played in human lives, weighing the evidence they offer and the practical implications for how to live and how best to pursue happiness.

IV

"ON THE HAPPY LIFE"

Seneca's Essay on the Happy Life

[W]hen once we have driven away all that excites or frightens us, there ensues unbroken tranquillity and enduring freedom; . . . there comes upon us first a boundless joy that is firm and unalterable, then peace and harmony of the soul and true greatness coupled with kindliness; for all ferocity is born from weakness.

Seneca, *De vita beata* (On the Happy Life)[1]

SENECA NEVER CEASED INVOKING THE STOIC VIEW THAT HAPPINESS required a simple, pared-down way of life in pursuit of wisdom and virtue. But as one who had accumulated great power and immense wealth in the service of the Roman Emperor Nero, he knew full well the criticisms of his opulent way of life.[2] "Yes, you claim to follow the Stoics. But how could you in all seriousness maintain, as they did, that virtue is all one needs to feel permanently happy? You, who have amassed such fortunes, first through marriage, then while holding public office in Rome? Who served as the tutor to Nero before he was proclaimed Emperor, and who remained as his counselor while he perpetrated some of his most horrific crimes? The very year, AD 59, when you wrote your essay on the happy life was the year Nero arranged for the murder of his own mother, Agrippina; some even said that you helped to contrive the boating 'accident' in which she perished. Exactly what could you mean, then, by invoking the Stoics on the subject of how to lead a happy life?"

In his essay, Seneca began his response to his critics on a personal note, addressing his older brother Gallio in defense of his way of life:

> To live happily, my brother Gallio, is the desire of all people, but their minds are blinded to a clear vision of just what it is that makes a life happy; and so far from it's being easy to attain the happy life, the more eagerly a man strives to reach it, the farther he recedes from it if he has made a mistake in the road; for when it leads in the opposite direction, his very speed will increase the distance that separates him.[3]

Seneca did not meet the severest criticisms head-on. To be sure, he admitted, he had not attained the status of a Stoic "sapiens," or wise man; he was not himself a fully virtuous person. "It is enough for me if every day I reduce the number of my vices and blame my mistakes."[4] Doing so required scrutinizing the difficulties bound to beset the pursuit of true happiness. "So long as we wander aimlessly, having no guide, and following only the noise and discordant cries of those who call us in different directions, life will be consumed in making mistakes . . ."[5] Instead we should ask what would bring lasting rather than temporary happiness; and for this, philosophers such as Socrates and the Stoic Zeno would provide the wisest guidance.

Once we set out on the right path with such guides, Seneca held, being rich need present no obstacle. On the contrary, wealth could bring legitimate pleasure, so long as it did not corrupt its owner. If asked whether he followed the Stoics, Seneca declared, he would say that he did so but that "I simply have this much to add."[6] What he had to add to their views was a more forgiving attitude towards the pleasures of life, including those that riches can bring to wise persons:

> As a favorable wind, sweeping him on, gladdens the sailor, as a bright day and a sunny spot in the midst of winter and cold give cheer, just so riches have their influence on the wise man and

bring him joy. And besides, who among wise men – I mean those of our school, who count virtue the sole good – denies that even those things which we call "indifferent" do have some inherent value, and that some are more desirable than others? To some of them we accord little honor, to others much. Do not, therefore, make a mistake – riches are among the more desirable things.[7]

Wealth and Other Goods

Still addressing his detractors in personal terms, Seneca drew a contrast: "If riches slip away, they will take from me nothing but themselves, while if they leave you, you will be dumbfounded, and you will feel that you have been robbed of your real self; in my eyes, riches have a certain place, in yours, they have the highest; *in fine*, I own my riches, yours own you."[8] For all that he called himself a Stoic, Seneca had much more in common with Aristotle's view that wealth gives people greater scope for exercising virtue. In speaking of the gifts of fortune that he had enjoyed and saw no reason to relinquish, Seneca found himself on Aristotle's side of the divide that V. J. McGill signals, between thinkers with expansive views about desires to be satisfied for achieving happiness and those convinced, like many Stoics, that it can be achieved only by having the fewest possible needs.

Today's findings on subjective well-being corroborate Seneca's claim that wealth can add to people's happiness so long as it does not dominate their lives or "own" them. As Ed Diener and Robert Biswas-Diener put it, "Yes, money buys happiness, but there are important exceptions." All in all, "rich people tend to be happier than poor people and people in wealthy nations are more satisfied with their lives than their less affluent counterparts."[9] This is hardly to be wondered at, given the advantages that money allows individuals to provide for themselves and for their families. Even lottery winners are happier than others, in spite of the publicity about winners of the highest sums. True, some find their lives turned upside down, with shopping sprees and binges bringing little satisfaction and relatives suing for a share of their gains, so that some end by facing bankruptcy. On average, however, people receiving small or medium windfalls and even larger sums report significant increases in happiness two years later. But the effect of money on felt happiness is often not large; and given that surveys report average figures, they do not apply to every individual: "We saw, for instance, how most homeless people were dissatisfied with their lives, but some were actually faring well in terms of happiness. And in the study of the Forbes list of millionaires, several were unhappy."[10]

Individuals who care most about acquiring wealth, who allow it to dominate their lives, who are owned by their riches, in Seneca's words, are, on average, less happy, less satisfied with their standards of living and with their lives than those who live by other values. But there is an exception: if those who value making a lot of money especially highly actually succeed in doing so, they do not tend to be less happy. Even people in this group, however, tend to pay less attention to family and friends than others; the satisfaction that they gain from wealth is derived at the expense of some of the joys that come from close relationships with others.[11]

Those who contribute money or time or effort to help others often experience increased happiness or satisfaction, confirming Seneca's view that peace and harmony of the soul come to those who exhibit kindliness. Seneca wrote more subtly, however, about the dangers of self-deception about one's own kindness and generosity than many who simply report the correlation between kind acts and higher levels of subjective well-being. In *De beneficiis* (On Benefits), he spoke at length about the complex issues of providing benefits to others and of receiving such benefits. He considered the effects of gratitude and ingratitude on the well-being of those who give and those who receive benefits; and held, with other Stoics, that beneficence is a subdivision of the virtue of justice that is its own reward regardless of whether it is reciprocated.[12]

This is one of many areas where I believe there should be a more fruitful dialogue between humanists and natural and social scientists. As the psychological studies of the role of money and riches accumulate, it will be possible to examine the factual assumptions by classical thinkers and to sort out those that are as valid today as in the past from those that should be set aside as, at least, not applicable to individuals today. Conversely, researchers may be able to refine their own inquiries by considering how Plato thought about wealth as one among other "instruments for use"; how Aristotle compared wealth and other "external goods" to, for example, "goods of the soul" such as moral and intellectual virtue; how Stoics debated the role of wealth; and Seneca's views on the effects on happiness of bestowing and receiving benefits.

What about Seneca's conception of the happy life as a life in harmony with nature, one that calls on us to have a sound mind, be courageous and energetic, be ready for every emergency, careful of the body and of all that concerns it, but without anxiety, and recognizing virtue as the only good? People who achieve such a state, he held, will live in constant cheerfulness and with a joy that is profound and issues from deep within, since they find delight in their own resources and desire no joys greater than their inner joys.

Social science research (along with everyday observation) takes issue with such a blanket assertion. Indeed, Seneca's contemporaries found such claims to be among the least probable assertions made by Stoics. It is easy to find people with a sound mind, courage, energy, and the other praiseworthy traits Seneca lists who nevertheless do not live in constant cheerfulness and joy, and are sometimes in deepest despair. Seneca himself must surely have encountered countless blameless victims reduced to such despair by Emperor Nero's bloodthirsty cruelties. Seneca's answer – the traditional one given by Stoics and one not persuasive to most challengers, then or now – was that there may be very few persons, perhaps none, who are sufficiently wise and virtuous to know full happiness.

Long before entering Nero's service and encountering nearly unimaginable cruelty and ferocity, Seneca had already expressed a dual vision of life as containing extremes of good and evil, glory and hardship. Writing to console Marcia, a mother so distraught after the death of her son that she asked how she could possibly go on living, he compared her question to asking whether or not to travel to a city such as Syracuse, in Sicily.[13] Just as that city had the balmiest of winters, a great harbor, many beauties, and much that would fill a visitor with wonder, so its inhabitants endured oppressive and unwholesome summers, and suffered under the tyrant Dionysus, "that destroyer of freedom, justice, and law, greedy of power even after knowing Plato."

> You will see man in his audacity leaving nothing untried, and you will yourself be both a spectator and a participant in mighty enterprises. You will learn and will teach the arts, of which some serve to maintain life, some to adorn it, and others to regulate it. But there, too, will be found a thousand plagues, banes of the body as well as of the mind, wars, robberies, poisons, shipwrecks, distempers of climate and of the body, untimely grief for those most dear and death – whether an easy one or only after pain and torture no one can tell. Now take counsel of yourself and weigh carefully the choice you would make; if you would reach these wonders, you must pass through these perils.[14]

Again and again, Seneca would return to asking what a happy life might be in a world where nothing is certain and where both wonders and perils offer choices between being "a spectator and a participant of mighty enterprises." In his own life, he tried to balance the two, as he wrestled with the question of whether to retire from active engagement if the circumstances were dire enough. It was not long after writing his essay on the happy life that it became a matter of life and death for him to do so. His choice would test his assertion that he would lose nothing of importance, nor be robbed of his real self, if his riches slipped away and if circumstances forced him to give up his life as "a participant in mighty enterprises." In AD 63, over sixty years old, he left the employ of Emperor Nero, more murderous and tyrannical every year, taking care to make over his fortune to Nero in the vain hope of placating him.

Once free, Seneca set about journeying from one location to another, corresponding with his friend Lucilius, reflecting on his search for the combined virtue and wisdom that would bring happiness, even in the face of impending death. Seneca brought into these letters dialogues that he had conducted throughout his life with thinkers such as Socrates, Aristotle, Zeno and Cicero, ending many of his epistles with a quotation from Epicurus on the value of living simply and of reducing one's needs. In Letter XXVII, for example, having declared that the only guarantee of lasting, carefree happiness is a good character, he closed with "Real wealth is poverty adjusted to the law of nature."[15]

It is hard not to conclude that Seneca, in this last period, lived up to the advice he had offered to his brother in "The Happy Life," to be the user, not the slave, of the gifts of fortune. In AD 65, he was having an evening meal with his wife Pompeia at his country villa outside Rome, when Nero's men surrounded the villa; a tribune accused him of plotting against the Emperor, and ordered him, on Nero's behalf, to commit suicide. Seneca became yet another example of someone who showed nobility in dying, and is often compared to Socrates, who was unjustly compelled to drink hemlock. With other Stoics, Seneca held that the doctrine of "reasonable departure" allowed individuals to choose to die when faced with insuperable assaults on their physical or moral integrity. The following year, his brother Gallio was likewise forced to take his own life – one more among the many victims of Nero's jealousy and paranoia.

The historian Arthur Darby Nock, writing in *Conversion* on philosophical doctrines and religions in the first centuries AD, contrasts two types of thinker: those willing to explore many different views but not fired with any desire to free humanity from error; and others, like Socrates, whose sense of mission finally leads them to martyrdom. Nock speaks of this distinction as one between *adhesion*, or acceptance of new beliefs as useful supplements, and *conversion*, the substitution of a new set of beliefs for past ones: the reorientation of the soul of the individual which is "seen at its fullest in the positive response of a man to the prophetic religions."[16]

Seneca was prepared to pick and choose from views on earthly happiness by thinkers such as Aristotle, Epicurus, and Cicero. But St Augustine, equally familiar with their teachings, struck an entirely different note after his conversion to Christianity. In his own essay on the happy life, he aimed to challenge pagan thinkers, along with all others who wrote about happiness in the here-and-now. The fall of man, recounted in Genesis, had destroyed all chances for true happiness on earth. It was possible only after death and only for those Christians who would come into the presence of God. Those who taught ways to achieve happiness on earth were leading their followers cruelly astray.

St Augustine's Essay on the Happy Life

Considering that the voyage to the port of philosophy – from which, indeed, one enters the hinterland of the happy life – must be charted only by rational choice, I do not think, great and noble Theodore, that I should be rash if I said that far fewer were likely to attain it than those scattered few whom we actually see reaching that port. For, since God or nature or necessity or our own will, or a combination of some or all of these, would have us founder in this world heedlessly and by chance as in a stormy sea . . . unless, at some time, a tempest . . . should thrust us, all unaware, off our faulty course upon the land so wished for.

Augustine, *De beata vita* (On the Happy Life)[17]

In the autumn of AD 386, only months after his conversion, Augustine composed an essay in the form of a dialogue on the subject of happiness, with almost the same title as Seneca's. He, like Seneca, used the metaphor of sailing, being thrown off course, and arriving into a safe harbor to introduce his view of happiness. It was a metaphor familiar to all who knew Homer's account of how Odysseus, shipwrecked and near death from drowning, had come to the island of the Phaiakians, who lived in peace and happiness. Theirs was neither the non-human happiness of the immortal gods on Olympus nor the everlasting existence of voluptuous ease that the nymph Kalypso had promised Odysseus if he stayed with her, but the fully human happiness with mortal delights such as those described by King Alkinoos:

And all our days we set great store by feasting,
harpers, and the grace of dancing choirs,
changes of dress, warm baths, and downy beds.[18]

Three years before writing "On the Happy Life," Augustine had taken his first sea voyage, from Carthage to Rome, and lived through a storm as terrifying as any he had read about in Homer or Virgil. This drama resonates in the first pages of his essay, as he relates what seafarers go through in attempting to reach "the port of philosophy," only to find that they do not know how to move from there to true happiness.

While composing his essay, Augustine was staying at a friend's country estate near Como to rest and recover from severe toothache and a painful throat disease.[19] In the limpid mountain air of Cassiciacum he hoped to pursue his inquiries with family members (including a recalcitrant older brother) and friends. For the first such occasion – one that would focus on happiness and that he described as "a feast for the soul" – he had chosen his thirty-third birthday, November 13, AD 386:

On the ides of November fell my birthday. After a breakfast light enough not to impede our powers of thinking, I asked all those of us who, not only that day but every day, were living together to have a congenial session in the bathing quarters, a quiet place fitting for the season. Assembled there were first, our mother, to whose merit, in my opinion, I owe everything that I live; my brother Navigius; Trygetius and Licentius, fellow citizens and my pupils; Lastidianus and Rusticus, relatives of mine, whom I did not wish to be absent, though they are not trained even in grammar, since I believe their common sense was needed for the difficult matter I was undertaking. Also my son, Adeodatus, the youngest of all, was with us, who promises great success, unless my love deceives me.[20]

All agreed that happiness was what everyone desires in life. Augustine, a superb dialectician, had no difficulty leading the group to accept three further propositions: that nobody can be happy without possessing what he desires; that everyone who is not happy is wretched; and that everyone, then, who does not possess what he wants is miserable.[21]

No one in the little group of converts questioned Augustine's assertions, however contrary to ordinary conceptions of what it is to feel happy or wretched; nor did they seek to qualify Augustine's statements about what "nobody" or "everyone" experiences in regard to happiness. They knew that he meant by happiness a state that is invulnerable to illness, death, or any misfortune. Such a state can only be hoped for in the afterlife. Augustine would therefore dismiss today's surveys of what people say about their state of happiness, satisfaction, or general well-being. The fact that many people who do not possess what they desire – say, good health or a thriving family – nevertheless say they are happy would carry no weight with him. Likewise, what people say about their present well-being would be beside the point, given his vision of the happiness that might be achieved after death – the possession of which lasts forever and is not dependent on fate "nor subject to any mishap." Happiness cannot, therefore, be anything "mortal and transitory." It can only be the happiness of possessing God, who alone is "eternal and ever remaining":

> Therefore, I concluded, "whoever possesses God is happy." As they readily and joyfully agreed to this, I continued: It seems to me, therefore, that we have only to inquire what man really possesses God, for he, certainly, will be happy. It is your opinion about this that I now ask.[22]

Licentius, pondering this question, suggested that he who lives an upright life possesses God. For Trygetius, it was one who does what God wills to be done. But the youngest member of the group, Augustine's son Adeodatus, held that "Whoever has a spirit free from uncleanness possesses God." When asked what that meant, he added that it involved living chastely.[23] Augustine's mother Monica, especially approved of the third suggestion, as did Augustine.

Intriguingly, had he observed that dictum earlier in life, he would not have lived outside of wedlock with the boy's mother and his beloved Adeodatus would never have seen the light of day.

Having inquired about whether living chastely did not mean to live free of all sins, and debated further about what seeking God and living righteously involved, Augustine offered his conclusion about the happy life, to which all assented: "This, then, is the full satisfaction of souls, this the happy life: to recognize piously and completely the One through whom you are led into the truth, the nature of the truth you enjoy, and the bond that connects you with the supreme measure."[24] Monica, who had played a central, revered role throughout the dialogue, then closed the discussion: "Indeed, this is undoubtedly the happy life, that is, the perfect life which we must assume that we can attain soon by a well-founded faith, a joyful hope, and an ardent love."[25]

Christian and Pagan Thinkers

In the background of the dialogue at Cassiciacum lay a much longer internal debate that Augustine had conducted before he was converted, with pagan thinkers such as Plato, Seneca, and above all others, with Cicero. Indeed, Augustine modeled his philosophical dialogues on those that Cicero had composed during his own retreat at Tusculum over three centuries earlier. But although Augustine took up pagan views such as those of Cicero, he did not invite questioning from anyone holding such opinions. Looking back in his *Confessions* on his early years as a teacher, he felt that he had been led astray at the time, and was leading others astray in turn. Now he aimed, with the zeal of a convert, to seek the sole truth and to lead others toward that truth, one that was to be found only in God.

After leaving Cassiciacum for Milan and throughout his long career as Bishop of Hippo, Augustine returned again and again in his writings to considering the human prospects for happiness; and he kept up sometimes vehement dialogues concerning the issues he had first raised as a young man, in sermons and letters and in discussions with friends and students. Much later, in *Retractationes*, he sternly reexamined much of what he had written, finding his first essays overly secular and urbane in spirit, still too influenced by those pagan thinkers he had studied so closely.[26]

Some later Roman Catholic thinkers drew more extensively on pagan philosophers. A hundred and fifty years after Augustine wrote his essay on the happy life, the eminent classical scholar and Christian thinker Boethius reflected on the possibility of happiness in his *Consolation of Philosophy*. But his was no festive discussion with family and friends at a country estate. He was languishing in a dungeon at Ticinum (the present Pavia, south of Milan) where, having been falsely accused of treason, he was awaiting execution. He, who had long known great happiness, was now so miserable and worn out that he longed for the delivery that death would bring. He decided to compose a dialogue in poetry and prose with *Philosophia*, the Lady Philosophy, describing her as having descended from on high to cure him of his despair, and recounting his misfortunes to her. What happiness could he possibly envisage after all that he had gone through? In the course of their exchanges, the Lady Philosophy gradually brought him to reaffirm the Augustinian view that although there was little hope for his or anyone else's happiness on earth, those who suffered misfortune nobly could hope for the perfect happiness of seeing God in the afterlife.

Boethius departed from Augustine, however, by praising Seneca as one who had been forced to choose his way of dying, identifying with him as one whose training in philosophy had made him seem "obnoxious to wicked men."[27] By contrast, Augustine had vehemently rejected the view of Seneca and other Stoics that suicide could be permissible under certain conditions. Declaring that the taking of one's own life was the one mortal sin that precluded every chance for subsequent repentance and divine forgiveness, Augustine became the first Christian to depict the torment in Hell after death as the punishment for self-slaughter – something that would contribute greatly to the strength of the prohibition on suicide in later centuries.

In the thirteenth century, Thomas Aquinas drew on the views of both Augustine and Aristotle – "the Philosopher" – in the analysis of happiness that formed a central portion of his *Summa Theologica*.[28] He rejected the various claims that happiness resided in factors such as wealth, fame, honors, power, health, or pleasure, citing Augustine and Boethius, though he agreed with Aristotle that each of these could indeed contribute to a happy life. Aquinas distinguished *beatitudo*, perfect happiness, from imperfect happiness, *felicitas*, and held that only the latter was possible on earth for those leading lives of virtue as Aristotle had defined it, along with the Christian virtues of faith, hope, and charity. And whereas earthly life is subject to many unavoidable evils, perfect happiness excludes every evil and fulfills every desire, for those who see God in the afterlife: "Final and perfect happiness can consist in nothing else than the vision of the Divine Essence."[29]

According to Aquinas, some among those fortunate enough to achieve such a state could relish it more than others, because of being "better disposed" to enjoy the vision of God.[30] Since there could be no eyewitness reports of the state of bliss in the presence of God, believers continued to debate its nature. Was it a passive state or one of active participation along Aristotelian lines? Just how did it differ from what could be experienced on earth? Did people, whether blessed or damned, retain their earthly bodies and physical characteristics? And were some individuals predestined to be able to enjoy such bliss or could they affect their chances through faith or through good works? During large parts of the sixteenth and seventeenth centuries, persecutions, inquisitions, and religious wars made the answers to such questions matters of life and death.

Reading Seneca: Elisabeth of Bohemia and René Descartes

In a letter to the brilliant and erudite Princess Elisabeth of Bohemia, in July 1645, René Descartes proposed that they both read Seneca's essay, *De Vita beata*. Then they could write one another about their reactions and discuss what philosophy teaches attaining happiness. One of the most useful ways to draw on philosophy in thinking of how to live, he suggested, was to examine what the ancients had written and "to try to go beyond them by adding something to their precepts; for in this way one can make them perfectly one's own and seek to put them into practice."[31]

Trained in the Jesuit tradition and familiar with every doctrinal controversy fueling the religious wars in Europe, Descartes lived in self-chosen exile in Holland, where he could enjoy relative freedom of thought and expression. Always careful to guard his writings against religious censorship, he knew the suspicions they aroused and would not have been surprised that they were placed on the Roman Catholic Index of Prohibited Books in 1663, thirteen years after his death. In corresponding with Elisabeth, a Protestant, it was natural, therefore, that he should have suggested they read Seneca's essay rather than what Augustine or Aquinas had written on the subject of happiness.

Their correspondence had begun two years earlier, when Descartes had received a letter from Princess Elisabeth that posed a question which, though formulated with the greatest humility, challenged the very core of his philosophy.[32] If, as he claimed, the spirit and the body were utterly distinct, and if the spirit, unlike the body, had no extension in space, then how can we do anything or say anything? How can our spirits influence our bodies, or our bodies our spirits in these or other ways? These were questions that Descartes continued to try to answer, both in his letters to Elisabeth and in his writings. He dedicated his *Principles of Philosophy* to her, saying that she was the only person he had ever met who could understand with equal ease what he wrote on mathematics and on metaphysics; and that, moreover, her magnanimity and gentleness and wisdom never altered, even in the face of injuries inflicted by fate. And his last work, *Les Passions de l'âme*, was "composed for the Princess Elisabeth."[33]

That their correspondence should come to focus on happiness, and above all on overcoming Elisabeth's often deep melancholy, was not to be wondered at. Descartes knew that her family had been driven into exile before she was two, when her father, then

King of Bohemia, lost his throne, and that he had died when she was thirteen. She was living with her mother and her seven siblings in the Hague, aware that the family's poverty precluded all marriage prospects for her. She ached to be able to devote more time to study; marriage, she insisted, mattered far less. She had studied history, scripture, the sciences, and the classics since childhood and was fluent, not only in Greek and Latin but also in French, English, Dutch, and German.

It was after Descartes learned that Elisabeth was in a state of profound dejection coming in the wake of lengthy bouts of fever and chest pain, that he proposed that the two of them study the theme of happiness together by reading Seneca. He claimed to have found the greatest possible happiness in the new life of seclusion and meditation that he had chosen for himself; and although he experienced periods of discouragement and melancholy about the slow progress of his work, he insisted that happiness was possible for all who are willing and able to rely wholeheartedly on reason to guide their lives.

In a subsequent letter, however, Descartes acknowledged that he had not in fact read Seneca's essay before proposing it to Elisabeth; he now found it wordy and too vague to offer practical guidance. To be sure, he accepted the fundamental distinction made by Seneca, following the Stoics, between things that depend on us, like virtue and wisdom, and those that don't at all, such as honors, riches, and health. Descartes agreed further that only the factors in the first category can bring beatitude and that they can do so in spite of any deprivation brought on by misfortune, poverty, or disease. He also went along with Seneca in rejecting the strictest tenets of some Stoics by admitting that those who are not only wise and virtuous but also favored by fortune have greater scope for contentment, indeed can experience more perfect contentment than those not thus favored. Not for him the claim that a thoroughly wise and virtuous person could be blissfully happy even while being tortured on the rack.

But just how are we to achieve the wisdom and virtue that alone bring genuine contentment? Seneca's advice to "follow Nature" helps us not at all. Instead, Descartes set forth three rules of morality as a guide for achieving contentment without depending on outside factors.[34] First, we should try to use our mind to know what we must do and not do in all the events of life. Second, we should have a firm resolution to carry out all that reason tells us to do without being led astray by passions or appetites. The firmness of this resolution, he added, should be thought of as virtue, though he knew no one who had explained it in this way. Third, we should accustom ourselves not to desire what lies beyond our power to attain. If Seneca had actually taught us to use our reason and to regulate our desires and passions so as to enjoy natural blessedness, his book would have been "the best and most useful a pagan philosopher could achieve."[35]

Elisabeth's answers to Descartes, in consecutive letters, stressed how much she felt she had learned from his analysis of Seneca's essay. But as for his three rules, she doubted that one could arrive at the beatitude of which he spoke merely by relying on reason and will. Certain maladies take away one's power of reasoning, while others reduce its force and make it impossible to follow maxims that good sense would otherwise dictate. Her own circumstances made it impossible for her to follow even that first rule, let alone the second concerning firm resolutions about not being led astray by passions and appetites.

Descartes backtracked right away, agreeing with her about illnesses that make it impossible to apply reason: "This teaches me that what I had said about all men ought to be understood as applying only to those who have a free use of their reason and who know the path one must follow to reach this beatitude."[36] As for conditions that do not completely trouble reason, but depend, rather, on "humors" or temperament and make people extraordinarily inclined to sadness, anger, or some other passion, these undoubtedly bring pain. But they can be surmounted. He firmly believed in people's ability to lift themselves out of melancholy, however severe, by their boot-straps. Senility or madness are one thing; but so long as there is the capacity to reason, he could see no problem.

Elisabeth persisted. Her descriptions of how illness, poverty, sometimes shame over misconduct by family members affect someone possessed of a melancholy disposition brought out the simplistic nature of Descartes's advice. And she added a crucial moral caution: Yes, happiness and contentment matter; but with all your talk about looking on the bright side of things, are we not likely to conceal our faults from ourselves – the very ones which, if we could look at them honestly, with whatever pain that would involve, could lead us to wish to change? Descartes responded obliquely, agreeing that it was more important to know even truths that make one sad than to be cheerful on the basis of self-deception. But if different considerations are equally true, "where some incline us to be content and others prevent us from being so," it is only prudent to dwell more upon those of the first kind.

That response cannot suffice, Elisabeth countered. Equal truth is far from all that matters. There may be equally true perceptions, one of which causes us greater regrets, because of our past actions, or greater pain in requiring us to bring about difficult changes in our lives. Virtue may require *not* avoiding troubling thoughts such as those about one's failings. And what about those persons who are too tormented by what befalls others to be able to feel happy? Our tendency to self-deception is so forceful that we set aside painful reflections about ourselves in ways detrimental to the moral life. Moreover, even if we tried to counter self-deception, just how should we go about strengthening our understanding so as better to discern what *is* best to do in all of life's actions?

Their exchanges ended after Descartes accepted an invitation from Queen Christina of Sweden to go to Stockholm and instruct her in philosophy. Shortly after arriving there, his last letter to Elisabeth spoke wistfully of longing to return to the solitude without which it was hard to pursue the studies that were his "principal good in this life." His brief stay in wintry Sweden turned out to be a miserable one: three months later, having had only a few early morning encounters with the Queen, and those quite dismal, he died from pneumonia, at fifty-three, in 1650. Elisabeth lived to a ripe old age, becoming Abbess of Herford, in Westphalia, continuing her philosophical correspondence with, among others, William Penn and, through her sister Louise, Abbess of Maubuisson in France, with Leibniz.

The correspondence between the two shows Elisabeth probing, pressing Descartes on metaphysical, moral, and psychological grounds; and Descartes, in turn, rethinking some of his positions. Yet in the end she held to her belief that happiness was a far more complicated matter than he thought, and that it was not always either simple or even advisable to try to set aside disturbing moral questions, even as he continued to defend his stance as someone who had found great contentment by setting aside moral concerns and sources of unhappiness that would interfere with his life's work.

Pascal's Wager

> But your happiness? Let us weigh up the gain and the loss involved in calling heads that God exists. Let us assess the two cases: if you win, you win everything, if you lose, you lose nothing. Do not hesitate then; wager that He does exist.
>
> Pascal, *Pensées*[37]

It was to atheists and skeptics more generally that the brilliant French mathematician Blaise Pascal directed his famous wager – a thought-experiment meant, like Nozick's Experience Machine, to challenge intuitive preconceptions about happiness. Pascal posited his wager in response to doubters asking why they should have faith in a deity of whose existence they saw no evidence. First of all, he used a mathematical analogy: just as we know that the infinite exists without knowing its nature, so we may know that God exists without knowing what He is. Second, he argued in strict cost-benefit terms that it would be worthwhile even for non-believers to seek to have faith, given the eternal happiness that is at least possible if God does exist. To the doubter's retort that the right thing would be not to wager at all, Pascal's response was peremptory: "Yes, but you must wager." Given this obligation to choose, not to wager that God exists is to wager that He does not:

"If you win, you win everything. If you lose, you lose nothing."
True, his belief that only a few elect souls favored by God's grace
will in fact achieve lasting happiness after death would seem to
undercut the certainty with which he predicted that the skeptic,
if accepting the wager, would "win everything"; but those who
refused to wager that God exists were bound to lose every slight
chance of such an outcome.

Though Pascal based his plea on claims about the chances of
eternal bliss or suffering in the afterlife, he added a final, practical
argument: that one who strove to believe in God, through prayer
and church attendance, would benefit in this life as well, by
becoming a better human being, more honest, humble, grateful,
and more capable of friendship. What's to lose, in that case, save a
variety of vices worth ridding oneself of anyway?[38]

Pascal would nevertheless dismiss as forcefully as Augustine the
relevance of today's psychological surveys of levels of subjective
well-being. And he would have nothing but scorn for the current
debates about whether religious faith more generally may
contribute to greater earthly happiness, better health, even longer
life. He ruled out all religious beliefs except his own strict,
Augustinian interpretation of the Christian faith; indeed, he wrote
with equal contempt against atheists and the deists who based their
belief on reason alone and gave little credence to supernatural
beliefs. No more than Augustine did Pascal believe that God had
intended earthly happiness for human beings, given that God had
punished all of humanity in perpetuity for Adam's lapse in Eden.

To Pascal, distress could be ennobling. Having endured
increasingly poor health from his late teens, he composed a heart-
felt prayer to thank God for having imposed illness and suffering
upon him as punishment for his sins.[39] Religion could therefore
never have been intended as preventive medicine. As for Bertrand
Russell's observation that belief in almost any cause, however
vacuous, is a source of happiness to large numbers of people,
Pascal would have found it symptomatic of blind carelessness
about the nature of true happiness and the infinitude of suffering
sure to befall non-believers in the afterlife.[40]

Earthly life, Pascal insisted, was utterly wretched; anyone who pointed to personal experiences of happiness lived in illusion. So did philosophers writing of how best to seek a happier life. "As men are not able to fight against death, misery, ignorance, they have taken into their heads, in order to be happy, not to think of them at all."[41] Among Christian authors, he saw most as guilty of demeaning errors and compromises by comparison with St Paul and Augustine. Pascal had converted to full acceptance of such beliefs during the night of November 23, 1654, at thirty-one years of age, recording his ecstatic assent in rising lines on a sheet of paper that he carried with him daily to the end of his life:

Certainty, certainty, emotion, joy, peace, God of Jesus Christ.
Deum meum et Deum vestrum.
Oblivion of the world and of everything except God.
Joy, Joy, Joy, tears of Joy![42]

In 1662, Pascal died after years of illness, not yet forty, leaving behind sheets on which he had noted the passages that would be published as his *Pensées*. Seneca and other Stoics, Jesuits, and philosophers – above all Descartes – were among those with whom Pascal took issue most passionately. Descartes, twenty-seven years his senior, had paid the young Pascal, already famous for having invented a calculating machine, a visit in his laboratory: an older scientist/philosopher coming to see a beginner already surpassing most others. But there had been no meeting of minds. Unlike Descartes, Pascal had concluded that reason had a role with respect to science but none at all when it came to religion. There, only faith could reveal what was true. All who, like Descartes, imagined that you could use reason to arrive at truths about God's existence and the afterlife were mired in error.

By the eighteenth century, Enlightenment thinkers increasingly rejected Pascal's views about the wretchedness of earthly life and the importance of thinking, above all, of true happiness as achievable only, if at all, in the afterlife. In 1734, Voltaire challenged Pascal in an essay that, while praising his genius, conjectured that he would surely have corrected his stern pronouncements, had he lived longer. "I dare to take the part of humanity against this sublime misanthrope; I dare to assert that we are neither as evil nor as unhappy as he says."[43] As for the wager, it was clearly false, Voltaire held, to imagine that those who do not bet that God exists go so far as to assert that He does not. The doubters' point should be that they cannot know enough to wager in one direction or the other. "I have an interest, to be sure, in there being a God; but if in your system God has come for so few people, if the small number of the elect is so terrifying, . . . do I not have a visible interest in being persuaded of the contrary?"[44]

Voltaire's essay met with resounding success. But only a few months after its publication, the French authorities declared it dangerous to religion and civil society; the book was burned and a writ was issued to arrest the author. Forewarned, he chose instead to flee Paris. Fifteen years later, his friend Denis Diderot's book *Pensées philosophiques* – his "anti-Pascal" – was similarly condemned by the Parlement de Paris as "scandalous and contrary to religion and good morals"; copies had been rounded up throughout the city and burned. Like Voltaire and other Enlightenment thinkers who had seen their books burned, Diderot took seriously the caution needed in the light of state and Church censorship. During the decades to come, he circulated only unpublished texts among friends, some even less diplomatic about religion than his earlier books. Taking a stand against obscurantism, he associated it, as did Voltaire, with the clergy of his day: "Lost in a vast forest at night, I have only a faint light to guide me. A stranger appears and says to me: My friend, you should blow out your candle in order to find your way more clearly. The stranger is a theologian."[45]

Against Pascal's wager, Diderot added a separate argument. Pascal had taken for granted that his wager concerned whether or not his particular view of God and of the afterlife were true or not. But the skeptic could hardly accept such a wager, which meant changing his entire way of life, given that other religions and beliefs were incompatible with Pascal's yet equally fervently held.

"Pascal has said: 'If your religion is false, you risk nothing in believing it to be true; if it is true, you risk everything in believing it to be false.' " To which Diderot simply responded: "An imam could have said as much."[46]

The entry on "Happiness" in Diderot's *Encyclopédie* refers to the shared views of philosophers that all people are united in seeking happiness; and asserts that the source of legitimate happiness flows no less for Christians than for non-believers; "but the Christian is infinitely happier from what he hopes than from what he possesses. The happiness he experiences here on earth becomes for him the seed of eternal happiness."[47]

Diderot and his encyclopedist colleagues might well, given what they took to be irreconcilable religious claims regarding what will bring true or perfect happiness, have agreed with the nineteenth-century British historian Alexander Kinglake's suggestion that the words "Important if true" should be posted over the door of all churches. But then they might perhaps add Salman Rushdie's comment, "Important even if not true."[48]

La Mettrie's Anti-Seneca

How anti-stoical we shall be! These philosophers are morose, severe, hard; we shall be joyous, soft, complaisant. All spirit, they abstract from their body; all body, we shall abstract from our spirit. They show themselves inaccessible to pleasure and pain; we shall glory in feeling both.

Julien La Mettrie, *Anti-Sénèque*[49]

Few have disagreed more vehemently with Seneca's, Augustine's, Descartes's, *and* Pascal's views on happiness collectively than the eighteenth-century physician and thoroughgoing hedonist, Julien Offray de La Mettrie. Having translated Seneca's essay on the happy life, he added an Introduction, longer than the essay itself, which offered his own views on ethics – the *Anti-Sénèque*, published anonymously in Potsdam in 1750, then reissued as *Traité du bonheur*. Published in 1747, his book *L'Homme-Machine* had caused a scandal by denying the Cartesian distinction between human beings, who alone had souls, and animals who, like machines, had none.[50] The book became known throughout Europe and its author reviled as "Monsieur Machine," an advocate of godlessness and vice. After the condemnation and burning of his book, La Mettrie was granted a safe haven by King Frederick II of Prussia and asked to serve as the King's personal physician.

Unlike other Enlightenment authors, La Mettrie expressly rejected all hope of immortality. Human beings, like all other living beings and the entire universe, he held, consisted of nothing but matter. If people are unhappy, it is because they experience remorse, inculcated through education, from which we should all liberate ourselves. Virtue was therefore not, as Seneca had held, a prerequisite for happiness. Sensation was all that counted for happiness, which was as possible for thoroughly selfish, greedy, even vicious, people as for other animals. A self-described voluptuary, La Mettrie advised reveling in every kind of pleasure. He would have denied that sensation inevitably wears off; what was needed, he held, was simply to become more sophisticated in orchestrating and varying one's pleasures.

For those not repelled by debauchery, La Mettrie famously suggested that "filth and infamy are there as your *glorious* allotment – wallow in it, as do the pigs, and you shall be happy in their manner."[51] He did not view such wallowing as desirable so much as understandable, for those who, because of their disposition and experience, find little satisfaction in "the highest arts of pleasure." Such people are more likely to drift into lives of vice and crime, unlike those who are equipped, as he felt he was, with enough imagination and understanding of voluptuousness to practice the subtlest forms of sensual delight of which he spoke. Let it not be said, he added, that he invited to crime; he only invited to "le repos dans le crime."[52]

In 1751, La Mettrie died unexpectedly at the age of forty-two – according to a rumor spread by Voltaire, after an attack of acute indigestion brought on by eating a whole pâté of pheasant with truffles. For one who had claimed to draw on his knowledge of medicine and who had vaunted his own good health as contributing to his happiness, accounts of the way in which he died seemed to indicate the errors of his teachings. Denis Diderot, who had known him since adolescence, described him as a "dissolute, impudent, clown, . . . [who] died as he had to die, victim of his intemperance and his folly: he killed himself through ignorance of what he professed." Diderot took special offense at La Mettrie's critique of Seneca, stating that it showed him to be "an author without judgment, who has spoken of Seneca's doctrine without knowing it," and who knows nothing about the true foundations of morality.[53] Diderot himself found Seneca's *De Vita beata* so beautiful that he would have liked to quote it almost in its entirety, summing up its message as "No happiness without virtue."[54]

When Descartes suggested that he and Elisabeth study Seneca's essay to see not only what they could learn from it but also add to it in thinking about how to achieve a happier life, he pointed to a path that Pascal, La Mettrie, Diderot, and countless others have taken, joining a long line of individuals seeking a dialogue with past thinkers on questions about what makes for a happy life. Augustine's essay on that subject is far less well known than Seneca's. Instead, countless readers have sought out Augustine's views in his *Confessions*, or turned for Seneca's views to his *Letters to Lucilius*. One reason, I believe, that so many readers have taken such vivid interest in the views of these thinkers is that they never hesitated to refer to their own life experiences in asking about practical steps toward the fullest happiness.

V

MEASUREMENT

Can Happiness Be Measured?

[T]he educated person seeks exactness in each area to the extent that the nature of the subject allows.

Aristotle, *Nicomachean Ethics*[1]

MANY PEOPLE MIGHT WELL AGREE WITH ARISTOTLE'S CAUTION against excessive specificity when it comes to assertions that happiness can be measured, treated in quantitative terms, dealt with by numbers and indices. Anyone claiming to have measured happiness is in for a series of questions. What measurements could possibly encompass the depth and scope of conceptions of happiness? By what quantitative standards can one person's happiness be compared to another's? What numbers make sense in comparing even one person's different experiences from day to day or year to year? And by what indices can such personal experiences be established by outsiders?

As for all who are convinced that true happiness can only be found in life after death, they would answer the question about whether happiness can be measured with a resounding collective "No," regardless of what conflicting views they might hold about immortality. By contrast, many social scientists now give an unqualified "Yes" to that question. We have worked out methods for measuring happiness, they claim, that use testable hypotheses and allow for greater specificity and exactness than anyone thought possible in Aristotle's time. Such measurements are not only possible; they are being carried out the world over. Three psychologists in the forefront of happiness research convey their confidence:

> In the last few decades there has been something of a revolution in the scientific study of happiness. A combination of radical new thinking and sophisticated methodology has allowed psychologists to add substantially to our understanding of this concept that has historically been the domain of philosophers and theologians. For the first time, we are able to measure happiness.[2]

Jeremy Bentham's Felicific Calculus

Over two centuries ago, the British jurist and philosopher Jeremy Bentham was equally certain that it was he who had shown for the first time how happiness could be measured by means of actual numbers in what he called his "felicific calculus." He saw no problem whatsoever in taking the measure of happiness with mathematical precision and gave short shrift to what ancient philosophers might have had to say on the subject. "It is not necessary to consult Plato nor Aristotle," he wrote. "*Pain* and *Pleasure* are what everybody feels to be such – the peasant and the prince, the unlearned as well as the philosopher."[3] Instead, Bentham drew for inspiration on eighteenth-century thinkers engaging in what Garry Wills has called "the numbering urge."[4] In France, the *philosophe* Helvétius described human beings as mechanistic calculators of pleasure and pain, happiness and unhappiness; the Marquis de Chastellux had proposed a science of measurable happiness, offering "indices of happiness" to explain the fate of societies; and the Bishop of Pouilli had even spoken of a thermometer of the spirit. Among Italians, the economist Pietro Verri had addressed all questions of happiness in terms of a calculation weighing desires and abilities, in his 1763 book *Meditazione sulla felicità*; a year later, Cesare Beccaria's *Dei delitti e delle pene* (On Crimes and Punishments) had astounded Europe by proposing ways of measuring and comparing forms of punishment, calculating for each one levels of certainty, intensity, and duration, and arguing for the abandonment of torture and capital punishment.[5]

Common to these thinkers and others at the time was some variant of the phrase "the greatest happiness of the greatest number," with Beccaria proposing that it should be the one criterion for evaluating all social policies and laws. That expression, Bentham wrote, made him feel converted for life when he first came upon it at the age of twenty, in a work by Joseph Priestley: "It was by that pamphlet and this phrase in it that my principles on the subject of morality, public and private, were determined."[6] He now had a system to guide all moral choice, all public policy. Like so many who offer novel definitions of happiness, Bentham redefined the concept drastically:

> By utility is meant that property in any object, whereby it tends to produce benefit, advantage, pleasure, good, or happiness (all this in the present case comes to the same thing) or (what comes again to the same thing) to prevent the happening of mischief, pain, evil, or unhappiness to the party whose interest is considered: if that party be the community in general, then the happiness of the community: if a particular individual, then the happiness of that individual.[7]

Just as Bentham held that "benefit, advantage, pleasure, good, or happiness" came to the same thing, as did "mischief, pain, evil or unhappiness," so he took their measurement to be simply a matter of "account and calculation, of profit and loss, just as [for] money."[8] All that was needed was to compute the value of each prospective pleasure or pain for an individual by measuring its intensity, its duration, the certainty or uncertainty of its coming about, its propinquity or remoteness, its fecundity (or likelihood of producing further pleasure or pain), and its purity (or "the chance it has of *not* being followed by sensations of the *opposite* kind: that is pains, if it be a pleasure: pleasures, if it be a pain"). One could then sum up these quantities and do the same for each person "whose interests appear to be concerned."[9] He even produced a rhymed formula to help people remember "these points, on which the whole fabric of morals and legislation may be seen to rest":

> *Intense, long, certain, speedy, fruitful, pure* –
> Such marks in *pleasures* and in *pains* endure.
> Such pleasures seek if *private* be thy end;
> If it be *public*, wide let them *extend*.
> Such *pains* avoid, whichever be thy view:
> If pains *must* come, let them extend to few.[10]

Bentham's proposals for how to understand and enhance human happiness were, in one sense, in keeping with Immanuel Kant's ringing assertion that the motto of Enlightenment is Horace's "*Sapere aude!*" (Dare to be wise!), which he rephrased as "Have courage to make use of your *own* understanding!"[11] But Kant could never have endorsed Bentham's beginning his first book, *A Fragment of Government*, by proclaiming that "it is the greatest happiness of the greatest number that is the measure of right and wrong."[12]

Throughout his long life, Bentham continued to insist that his own felicific calculus brought even greater quantitative specificity of measurement than those of his predecessors. Relying on that calculus, he worked ceaselessly at proposing ways to improve not only individual conduct but all public policy concerning governance, education, and laws.[13] He was one of the first to speak out for the equal rights of women, including the rights to vote and to be divorced; and he advocated the end of slavery and the abolishtion of physical punishment, including the brutal treatment of children in the name of character-building. He also championed the rights of animals, arguing that it was their ability to suffer, not their ability to reason, that should govern their treatment. His grand project was the Panopticon prison, a circular building he proposed to build and to run, in which an overseer would sit at the center, unseen by the prison inmates but observing their every move. He proposed a similar plan to deal with poverty, in which poor people would be housed, fed, and clothed in a system of workhouses around a central observation post, then put to productive labor.

By combining an all-inclusive definition of happiness, however, with a spuriously specific method of quantitative measurement to resolve all questions of right and wrong, and with faith in an overriding moral obligation to "minister to general happiness," Bentham invited royal chaos with regard to all three. The felicific calculus, he declared, was available for all persons surveying the effects that their actions could have on all interested parties. These persons alone were to estimate the expected effects, for themselves and others. There was no need to consider what the "interested parties" might have to say about the estimates agents made of their pleasures and pains. And only the utility of the fore-seeable consequences of the agents' actions should count for the purposes of such estimates, rather than intentions, character, or the adherence to any moral principle other than Bentham's own Greatest Happiness Principle.

Critics have rightly challenged Bentham on all these scores. Why should people trust themselves or anyone else to perform such computations with mathematical precision? Even if agents could somehow predict the effects of different choices on their own pleasure and pain, why should they assume that they could also know the effects upon other persons – let alone upon "the happiness of the community"? And, given the priority for most people of their own happiness and that of those close to them, why should they be expected to choose the action providing the greatest happiness of the greatest number, if that meant sacrificing their own?

By the time Bentham labored over his last work on ethics, the *Deontology* that remained unpublished at his death, at age eighty-four, in 1832, he still hoped that his calculus could shed light on individual and societal choice. Neither his reformist zeal nor his claims to have answers to all moral problems had diminished. Quite the contrary. The title page proudly proclaimed his aim:

Deontology:
or
Morality made easy
Shewing how
Throughout the course of every person's life
Duty coincides with interest rightly understood
Felicity with Virtue
Prudence extra-regarding as well as self-regarding with
Effective benevolence

Bentham had nevertheless tempered his views about both the definition and the measurement of happiness. The term "happiness," he now conceded, made it hard to sum up pleasures and pains: "it seems to leave pains out of account and concentrate on pleasures experienced in a high and, as it were, superlative degree."[14] Not many people enjoy such happiness over the whole length of their lives, whereas most enjoy "a measure more or less considerable of well-being." The concept of "well-being" lends itself more readily than "happiness" to calculating "the sum of pleasures of all sorts, and the sum of pains of all sorts which down to the point of time (suppose the end of his life) a man has experienced." It was to be hoped, he added, as if acknowledging shortcomings of his felicific calculus, that a way would one day be worked out to make such calculations more accurate.

Again and again Bentham returned to the question of why people performing the calculus should consider the consequences for the general happiness rather than only their own. He recognized that this was clearly a challenge he had to meet if his entire system was not to collapse. How could he maintain that we both do and should seek always to increase our own happiness, yet that we should take into account, in all our choices, the effects on all those affected by our actions?[15] He eventually stressed the role of sympathy that he had earlier downplayed: we humans, he claimed, are endowed with sympathy for the pains and pleasures of others and with love of reputation, so that making others happy increases our own happiness while being known to do so adds the utility of relishing our greater worth in the eyes of others. True, some individuals seem capable of deriving felicity from vice, cruelty and crime. It was for them that he saw what he called "the great engines of punishment and reward" as most needed. But most others would naturally seek the general happiness, he could still declare; they would have to make no corresponding sacrifice of their own, so long as they were made happier by their contribution to the happiness of others.

John Stuart Mill: How Measure Both Quality and Quantity?

"A boy to the last" – such was John Stuart Mill's judgment of the man whose fervent disciple he had once been and whose utilitarian principles he still endorsed. After Bentham's death, he wrote to reassess Bentham's role as a thinker and reformer, both praising him as a "great questioner of things established" and father of English innovation, and asserting that his inability to understand the thought and circumstances of other individuals left him blind to points of view different from his own. Because Bentham dismissed as "vague generalities" all thinking not founded on a recognition of utility as the moral standard, "the nature of his mind prevented it from occurring to him, that these generalities contained the whole unanalyzed experience of the human race."[16]

By setting aside the views of thinkers who disagreed with him, Bentham cut himself off from the dialogues with past thinkers, such as Aristotle, Plato, Montaigne, and Kant, that Mill himself found indispensable.[17] Symptomatic of Bentham's narrowness was his attitude toward poetry as "misrepresentation" and his insistence that any pleasure poetry afforded was indistinguishable from a comparable pleasure from the game of "pushpin." By contrast, it was poetry, Mill explained in his *Autobiography*, that had saved him from overpowering despair after he suffered a breakdown in his late teens. Having been reared in the Benthamite creed to believe that his happiness would come from being "a reformer of the world," the collapse came when he asked himself whether it would be a great joy and happiness to him if all the desired reforms were to succeed:

"And an irrepressible self-consciousness distinctly answered, 'No!' At this my heart sank within me; the whole foundation on which my life was founded fell down." Poetry was an important factor in his recovery, and especially reading Wordsworth, in whose poems he felt he drew on a source of inward joy, of sympathetic and imaginative pleasure: "From them I seemed to learn what would be the perennial sources of happiness, when all the greater evils of life shall have been removed. And I felt myself at once better and happier once I came under their influence." Feeling was to be at least as valuable in his version of utilitarianism as the powers of reasoning and analysis that he had been so relentlessly trained to employ; and poetry, he now saw, provided the necessary condition of any true and comprehensive philosophy.[18]

Mill still accepted Bentham's Greatest Happiness Principle as the foundation of morals and fully agreed that "the truths of arithmetic are applicable to the valuation of happiness."[19] But he found Bentham's felicific calculus too rudimentary to provide accurate measurements. Without including the qualitative differences between forms of pleasure, any calculus would be worthless. Mill referred to Epicurus and past thinkers who had taught that although happiness was the end of life they did not mean a life of rapture, "but moments of such, in an existence made up of few and transitory pains, with a decided predominance of the active over the passive."[20]

How, in that case, incorporate in the measurement more than the simple quantities of pleasures and pains to which Bentham had reduced all human happiness and unhappiness? How include, for example, pleasures from reading or enjoying close human relationships or the uses of one's imagination that cannot possibly be broken down into segments to be measured by the criteria of the felicific calculus? To carry out such measurements, Mill substituted a scheme that included preferences regarding higher and lower pleasures for Bentham's rudimentary calculations of pleasures and pains. In so doing, Mill abandoned a central claim of Bentham's: that the calculus would be simple to carry out for "the unlearned as well as the philosopher." Mill chose, instead, the path of Aristotle, holding that people were not all equally qualified to make such estimates. Some are more developed, emotionally and intellectually; they have more experience of the higher pleasures, not only of the lower ones. They are therefore better able to carry out the measuring accurately and wisely, first taking into account Bentham's criteria of quantity, duration, intensity, etc., and then adding on quality.[21] "The test of quality, and the rule for measuring it against quantity" should be "the preference of those, who in their opportunities of experience, to which must be added their habits of self-consciousness and self-observation, are best furnished with the means of comparison."

Whereas, for Bentham, the quantitative summing up brought indisputable evidence, Mill proposed that it would be the judgment of the superior persons themselves that would determine how to compare preferences for higher and lower pleasures: "From this verdict of the only competent judges, I apprehend that there can be no appeal."[22] In granting ultimate authority to their decisions, however, Mill still failed to address the problems of interpersonal comparison that beset Bentham's scheme. Why should we trust any person, wise or not, to be able to make accurate estimates about the pleasures and pains of other people? Just why, moreover, should the preferences of "competent judges" be given priority? And what risks does such prioritizing present for others deemed less competent?*

Even though he hewed to the Aristotelian view of the judgment of the educated few as superior to that of the many, Mill took issue

---

* To cut back on such risks, Mill argues, in *On Liberty*, against interfering with "self-regarding" choices, however ill-advised, that individuals make, so long as these choices do not affect the interests of others. But because he offered no clear ways of discerning just when the interests of others might be affected by a choice, this argument provides little protection, even in principle, against coercion. And those most clearly exposed to coercive interference are at the opposite end of the "development," mental and moral, that Mill prized – young people and states of society he saw as so undeveloped as to be in a state of "nonage." As for those he characterized as "barbarians," despotism was a legitimate mode of government in dealing with them, "provided the end be their improvement." (Mill, "On Liberty," in Warnock, *John Stuart Mill*, 135–6.)

with Aristotle on another score: one that scandalized most of his contemporaries. He insisted that women were as capable of achieving such superiority as men. In *The Subjection of Women*, he sided with Bentham in pointing to the immorality and the absurdity of "the legal subordination of one sex to the other."[23] With his wife, Harriet Taylor, he had fought against laws such as those governing marriage, legal status, voting, and work. They both saw the servitude imposed on women as blocking off the educational opportunities open to men that would promote their emotional and intellectual maturation, with the result that many women experienced the very opposite of happiness: "weariness, disappointment, and profound dissatisfaction with life." The book concluded by stressing how the restraint on women's freedom of conduct dries up "the principal fountain of human happiness and leaves the species less rich, to an inappreciable degree, in all that makes life valuable to the individual human being."[24]

Their words ring as true today as a century and a half ago. So does their utilitarian argument that depriving women of freedom "dries up the principal fountain of human happiness" and thereby leaves everyone – "the entire species" – less rich in all that makes life valuable for individuals. Their argument is sadly corroborated by comparisons between societies that have moved ahead with the reforms the authors advocated and societies that still reject such reforms, including access for girls even to elementary education, at times so savagely combated.

Francis Edgeworth and the Prospect of Happiness Meters

In 1881, the young barrister and political economist Francis Edgeworth published a slim volume entitled *Mathematical Psychics: An Essay on the Application of Mathematics to the Moral Sciences*. He wholeheartedly endorsed the utilitarian Greatest Happiness Principle; but while professing the deepest respect for Bentham and Mill, he aimed to correct some of their conclusions and offer a more powerful and mathematically precise "hedonic calculus" to guide all moral, political, and legal choice.[25] He criticized "the great

Bentham" for having adopted the meaningless formula "the greatest happiness of the greatest number" as the creed of his life and the watchword of his party. The expression was meant to be quantitatively precise, yet was it really more intelligible than "greatest illumination with the greatest number of lamps"?[26]

Edgeworth explained that it was fully possible to compare estimates of quantities of pleasure without adding and subtracting numbers, as Bentham had, by relying instead on "data which, while not numerical, are nevertheless quantitative." A man can make such a quantitative estimate of a pleasure unit at a particular time, then calculate "just perceivable increments" to this pleasure. As for calculating such "pleasure units" of other sentient beings, these could only be inferred, he admitted; but the uncertainty thus brought into the calculations would be compensated for by the greater number of measurements taken. Accordingly, he added, venturing into sheer speculation, "We have only to add another dimension expressing the number of sentients and to integrate through all time and over all sentience, to constitute the end of pure utilitarianism."[27]

Edgeworth further postulated that "any just perceivable pleasure-increment experienced by any sentient at any time has the same value." But because individuals and groups vary in their capacity for pleasure, he took Mill to have been wrong in holding that every person should count for one, and no one for more than one.[28] There was nothing in the Greatest Happiness Principle that required such equality, Edgeworth insisted. Rather, persons and groups possessing the greatest capacity for happiness should be granted precedence. Whenever that capacity is equal, then both happiness itself and the means thereto ought to be distributed equally; "if unequal, neither". And since he believed that capacity varies between the sexes and according to levels of wealth, civilization, and education, just trying to distribute happiness equally will achieve neither the goal of equality nor that of the greatest amount of happiness. It is "repugnant to the Utilitarian End."

Studded with mathematical formulae and relying on intuitive, unsupported premises about different capacities for pleasure, the book championed discriminatory policies that Bentham, Taylor, and Mill would have rejected as morally unacceptable. Underscoring what he called his Aristotelian preferences, Edgeworth pointed to a sentiment shared by "many good men of the moderns and the wisest among the ancients . . . in favour of aristocratic privilege – the privilege of man above brute, of civilized above savage, of birth, of talent, and of the male sex."[29] He waxed poetic on the subject of women's inferiority, referring to men's greater *energeia* and superior capacity for happiness:

> Woman is the lesser man, and her passions unto mine
> Are as moonlight unto sunlight and as water unto wine.[30]

If his sharp-witted aunt, the Victorian writer Maria Edgeworth (whose *Popular Tales* John Stuart Mill remembered having enjoyed as a child, alongside the *Arabian Nights* and *Don Quixote*) had lived long enough to come across such syrupy condescension toward women on the part of her nephew, she would surely have taken him to task. But his elitist sense of privilege extended to all persons and groups that he thought inferior. "In fact, the happiness of some of the lower classes may be sacrificed to that of the higher classes," Edgeworth had stated in an earlier essay. "To those who have, more shall be given," so long as the poorest remain above the starving point; famine would damage the utility for all. Political institutions should be changed accordingly.[31]

To show how far quantitative measurement might one day extend, Edgeworth brought in the metaphor of a hedonimeter, an imagined "psychophysical machine," for calculating experienced units of happiness over time with the greatest possible precision. From moment to moment, the hedonimeter would vary, "the delicate index now flickering with the flutter of the passions, now steadied by intellectual activity, low sunk whole hours in the neighbourhood of zero, or momentarily springing up towards infinity. The continually indicated height is registered by a photographic or otherwise frictionless apparatus upon a uniformly moving vertical plane."[32]

Today's neuroscientists have recourse to remarkable new methods for measuring intensities of pleasure and pain and many other brain responses, including actual probes sunk into the brain capable of measuring just the kinds of fluctuating responses Edgeworth had imagined. By using methods of "deep brain stimulation," sending electrical pulses to specific sites in the brain, neurosurgeons are able to relieve the suffering of patients with Parkinson's disease; and research is under way to use such stimulation to alleviate states of profound sadness among patients with intractable depression.[33]

But while the new methods give credence to some of the possibilities for measurement that Edgeworth envisaged, they assuredly do not confirm his speculations about race and gender as determining which groups are more capable of pleasure than others. As with Bentham's felicific calculus and Mill's revision of it, the problem with Edgeworth's "hedonimetry" was not the effort to *measure* pleasure and pain but the uses to which he meant to put his method. Given his reliance on unsupported assumptions, both about being able to measure happiness more generally and about exactly whose evaluations mattered, there was little wonder that he would arrive at unwarranted conclusions about individual well-being and social policy.

The sophisticated new methods for measuring brain activity, moreover, offer no support for Edgeworth's belief that a machine such as his would be capable of calculating "units of happiness" that would satisfy even the most bare-bones definitions of happiness, let alone those of Aristotle or Kant. Indeed, the new techniques challenge any simple "hedonic calculus" in which pleasure and pain lie along one dimension. It turns out that there are distinct and independent neurological substrates for many forms of positive and negative feelings.[34] Neurologist Antonio Damasio, using brain scanning techniques to help map the anatomy of the human brain, persuasively points to a far more complex role of pleasure and pain. He sees these as the essential ingredients of all human feelings, underlying the vast variety of feelings and emotions, and forming "the continuous musical line of our minds, the unstoppable humming of the most universal of melodies that only dies down when we go to sleep, a humming that turns into all-out singing when we are occupied by joy, or a mournful requiem when sorrow takes over."[35]

Economists have long since abandoned Bentham's cardinal measurements of pleasure and pain, opting instead for ordinal measurements of individuals' needs on scales of preferences about, for example, consumer goods or housing arrangements. The problem of comparing utilities among individuals has remained, however. Even economists who think it possible to attach some proximate numerical values to individual preferences recognize the difficulties in arriving at interpersonal comparisons of this kind.[36] But whether they stress subjective preferences or other evaluations of well-being, many have tended to pay minimal attention to empirical studies of what human beings actually say about their experience of well-being and happiness.[37] By now, however, many economists, not least those in the field of behavioral economics, are once again engaging in the kinds of interdisciplinary cooperation that earlier thinkers such as Adam Smith and Mill took for granted.

Psychological Studies

For over fifty years, social scientists have worked to improve methods for studying people's views of their own subjective well-being. Most now take into account three components of such well-being. Two are empirical in nature: high levels of "positive affect"; and low levels of "negative affect." The third is cognitive. It involves people's evaluation of how satisfied they are with their lives, and is only moderately correlated with the first two.[38] Excruciating experiences such as life-saving surgery can be accompanied by a sense of high satisfaction with one's life, just as the delights of overeating may go along with rueful self-evaluation. Earlier work, primitive by today's standards, still relied on intuitive assumptions about who was and was not happy. In 1967, for example, Warner Wilson carried out the first broad review of "Correlates of Avowed Happiness."[39] His conclusions were that the happy person is a "young, healthy, well-educated, well-paid, extroverted, optimistic, worry-free, religious, married person with high self-esteem, job morale, modest aspirations, of either sex and of a wide range of intelligence." The image projected, of what might be a 1960s' stereotype of a relentlessly cheery, self-confident, and somehow worry-free young person, is almost as dismaying, as a model for happiness, as that of Edgeworth's wealthy, privileged, highly "civilized" male.

Nine years later, an ambitious, more carefully calibrated study aimed to complement objective measures of factors such as population growth, income, health, housing, employment, marital status, and levels of education in America: "The fact is that we do not know how well objective measures such as these represent underlying psychological states or how well social indicators can be taken to present the quality of life experience."[40] It mattered, the authors stressed, to go beyond such objective factors and to investigate, also, less tangible dimensions of people's satisfaction with their lives, such as the ways in which they evaluated their work experience, their marriage and family life, their residential environment, and their communities, as well as differences in the quality of life experience of women and minorities.

By the late 1980s, the research on happiness, well-being, and life satisfaction was coming into its own. Much of the credit must be given to two social scientists: in America, Ed Diener, who published his first article on subjective well-being research in 1984, and in the Netherlands, Ruut Veenhoven, author of *The Conditions of Happiness* which appeared the same year. Both of them went on to do research on many aspects of the topic, often with family members and colleagues, and, increasingly, with sociologists, economists, and biological scientists. By now, we are seeing worldwide collaboration in the field of happiness studies on a scale never before undertaken, providing measurements comparing individuals, communities, and nations on a vast variety of scores.[41]

The new research aims to provide testable hypotheses and conduct studies using a range of different measurement methods, so as to control for errors in any one of them. For example, psychologists conduct randomized experiments, dividing subjects into two groups, then asking those in one group but not the others to perform a particular task, then measure levels of happiness in both. The expectation is that they are randomly distributed, except for the experimental task. For instance, researchers may ask one group of students to engage in specified acts of kindness, then measure any change in their level of subjective well-being as compared to members of the other group. The drawback with such studies is that they are seldom conducted over long periods of time, so that it is hard to know how long the measured changes persist.

A frequently used method is the "Satisfaction with Life Scale," pioneered by Diener. It asks individuals to indicate whether they strongly agree, agree, slightly agree, neither agree nor disagree, slightly disagree or strongly disagree with these statements: "In most ways my life is close to ideal." "The conditions of my life are excellent." "I am satisfied with my life." "So far, I have gotten the important things I want in life." "If I could live my life over, I would change almost nothing."[42] Ruut Veenhoven describes a related method, asking people in many societies a single question: "How satisfied are you with your life-as-a-whole?"[43] The findings from these methods can then be compared to what family members, friends, and colleagues of the respondents have to say about the levels of happiness, and to surveys of the same subjects over longer periods of time. True, outsiders may regard people's satisfaction with their own lives as off-base, even deluded, yet still agree about these individuals' levels of *felt* satisfaction.

Psychologist Daniel Kahneman and colleagues have carried out two kinds of studies to measure well-being "anchored in the reality of present experience, not in fallible reconstructions of and evaluations of the past." The first known as the "Experience Sampling Method," supplies people with a Palm Pilot or other electronic diary that beeps at random times during the day, then inquires what they were doing just before the prompt and what intensity of feelings such as happiness and frustration they were experiencing just then. The second type of study, the "Day

Reconstruction Method," asks participants to fill out a diary about events during the previous day and answer questions about experiences during that day.[44]

For individuals reflecting on their own experience, it is helpful to imagine taking part in both kinds of survey. Most will surely find wide variations in the answers they might give depending on mood and circumstances. Say that the buzzer sounds for the Experience Sampling survey when you are taking a beach walk at sunset, seeing pelicans diving for fish; then sounds again while you watch the evening news, witnessing the plight of families fleeing forest fires or floods. The surveys show similar individual variations, and calculations of average differences in subjective well-being for different activities such as social encounters, doctor's visits, sexual intercourse. Commuting to work turns out to be especially low on most people's scale of reported well-being.

We can also easily imagine taking part in the Day Reconstruction Method, since it is so similar to keeping a journal or simply to looking back at the end of the day to see what one has done and experienced, perhaps learned. It can be of personal use to think of participating in such a survey, in two ways. Doing so may shed light, first of all, on one's preferences and on any differences between encounters and other experiences considered before and after they took place. Second, looking back may reveal a moral dimension not present, or even appropriate, in the psychological questionnaires used in the surveys. In many philosophical and religious traditions, such reflection at day's end has more to do with moral questioning than with levels of happiness and satisfaction, as the term "examination of conscience" indicates. For example Pythagoras, the sixth-century BC philosopher and mathematician, is said to have asked that his followers take time, before going to sleep each night, to pose three questions: Where have I failed myself? What have I done? What duty have I not fulfilled?[45]

Individuals reflect on moral questions for their own purposes while researchers and experts are extracting data for other aims, and have other priorities. We surely have much to learn from how large numbers of people evaluate their experience. Taken together, the findings from the collaborative studies by natural and social scientists that are now accumulating represent a great improvement over past guesswork and slapdash intuitive comparisons of individuals. Such comparisons make it possible to disprove anecdotal, often erroneous intuitive claims, as when John Stuart Mill declared, in *Utilitarianism*, that it was unquestionably possible to do without happiness since it was "done involuntarily by nineteen twentieths of humanity."[46] The new findings drawing on large samples of people from different walks of life and different societies also challenge a number of the moral and political assumptions about happiness by past thinkers, such as Edgeworth's supposition about men's and aristocrats' inherent superior capacity for happiness. As for the common belief first voiced by Aristotle, that people become grumpier and more unhappy as they grow old, this is definitely shown no longer to be the case, if it ever was. Among equally healthy people, those over sixty-five are more satisfied with their lives. But because older people are, on average, less healthy, that cuts into their subjective well-being. Likewise, more recent findings turn out to have been premature, such as those in the 1967 study by Wilson, mentioned above, to the effect that not only youth but also good education and modest aspirations correlate strongly with felt happiness.

Important questions nevertheless remain about the validity of the findings regarding subjective well-being and happiness more generally. Many researchers would agree that there is reason to be cautious in the face of the proliferating claims about happiness that are said to be based on scientific measurements. These are, after all, not measurements of happiness *per se*, but of brain responses, physiological reactions, performance on psychological experiments, and answers to survey questions.

Findings from such methods can be skewed in several ways. Different types of measurement show different levels of satisfaction, and responses can be influenced by factors such as having witnessed a cheerful or depressing film clip. Results can also vary depending on the groups surveyed, to the point that it may be difficult to draw general conclusions from them about particular correlations. How can we know the degree to which questions and answers about well-being and satisfaction in diverse languages and cultures are comparable given the wide differences in how happiness is defined and in value judgments regarding what matters for human lives to be good ones?[47]

A case in point has to do with the research on correlations between nationality or religion and levels of happiness. For instance, while studies of Americans tend to show modest correlation between regular attendance at religious services and happiness, the same is not the case with respect to Scandinavians and the Dutch, who are often cited as higher on scales of subjective well-being than Americans.[48] It matters to ask about the types of community support and fellowship that citizens in the different countries enjoy and whether Americans are more likely than others to seek such support through religious activities. Likewise, it is important to include, in surveys to do with religious belief, questions addressing the powerful influence of hopes and fears about what might befall individuals in the afterlife.

Then there is the complexity regarding the direction of causation. Take the correlations found between felt happiness and marriage. Does marriage contribute to such happiness or are people with higher levels of subjective well-being more likely to get married in the first place? The correlation between health and felt happiness likewise raises immediate questions about the direction of causation. Many findings indicate that those who are healthier are more likely to be happy; but how can one know whether happiness serves to increase the likelihood of good health in the first place? Health has long been considered as essential to happiness. Some have even made *perfect* health a defining characteristic of happiness: one without which no one could possibly be happy. An eighteenth-century French author, the Marquis d'Argens, specifies in "On the Happy Life" that true happiness requires three things: "not to have anything criminal to reproach oneself for; knowing how to make oneself happy in the state where Heaven has placed us and where we are obliged to remain; and enjoying perfect health."[49] Yet there is much evidence to refute each of the three criteria, most specifically the requirement of perfect health. Most people are happy or moderately happy though few, if any, can boast of perfect health; and a number of people testify to a heightened sense of happiness after being diagnosed with cancer or some other life-threatening disease.

An additional factor complicates the correlation between health and felt happiness. A high correlation exists between how people view their own state of health and how satisfied they say they are with their lives more generally; but it is much lower between such satisfaction and what physicians regard as their actual health. This is only to be expected, given how differently people deal with their health problems, with some adapting with greater resilience to what weighs others down, and some worrying far more about even minor health issues than others. The very fact of believing oneself in good health in spite of evidence to the contrary, some researchers suggest, helps to prolong life, to the extent that it produces a higher level of subjective well-being.[50] On the whole, perceived health and actual health turn out to be weakly correlated; and neither is found to be indispensable for happiness.

The same is the case with perceived moral excellence, which may enhance a sense of self-respect and satisfaction, no matter how delusory. Imagined generosity, courage, wisdom, intelligence may contribute to self-esteem and to subjective well-being while utterly at odds with outsiders' evaluations. This is especially likely to be the case for characteristics that are subjective in the first place and cannot easily be measured. On this score, such traits differ from, for example, the imagined belief that one is a great pole-vaulter or violinist.

Neither health nor any other factor or set of factors turns out to be truly *necessary* for people to feel happy or satisfied with their lives. Human beings are so complex, and experience happiness in so many different ways, that for any factor believed to be indispensable, a multitude of examples reveal people's ability to thrive without it.[51] Yet while no single factor contributes to subjective well-being for everyone, most do so for some people and a few factors, such as being free of profound depression or chronic pain, do so for almost all. So-called demographic factors, such as gender, age, and ethnicity, that have too often been thought determinative of the very possibility of happiness, turn out to play a less significant role in modern democratic societies. In this respect, such societies clearly differ from those still retaining the most brutal forms of institutionalized religious, gender, ethnic, or other discrimination. The factor found to be most important for subjective well-being is that of social relationships with family, friends, and others. Other factors that correlate significantly with subjective well-being are income or wealth, personality traits such as optimism and extroversion, and occupation or employment status.

In response to the variety of methodological challenges, psychologists Oishi, Diener, and colleagues underscore the importance of combining different methods of measurement, including self-reports, peer reports, observational methods, longitudinal studies, experimental studies, and physiological and other methods so as to refine the interpretation of responses to surveys; and of doing so before drawing general conclusions, even for a single society, let alone many, wherever possible.[52] And Daniel Gilbert brings to bear "the law of large numbers" as a partial answer to legitimate concerns about the difficulties of measuring subjective experience. When a measurement is too imperfect for our tastes, he argues, we should not stop measuring, but measure again and again until niggling imperfections yield to the onslaught of data. True, Gilbert adds, the study of what cannot be quantified – what he calls "unquantifiables" – might arguably be more worthwhile than all the sciences laid end to end: "But it is not science because science is about measurement, and if a thing cannot be measured – cannot be compared with a clock or a ruler or something other than itself – it is not a potential object of scientific inquiry."[53]

What about the "unquantifiables" of which Gilbert speaks? Few of the experiences of happiness that are conveyed in autobiographical writings and literature can be fully measured by psychological or neuroscientific research. Nor can most of the philosophical and religious claims about the nature of happiness or about the role it plays in human lives. Must such issues, then, be thought of as lying beyond the purview of what has come to be called "the science of happiness"?

A Science of Happiness?

Many recent books on happiness refer to "The Science of Happiness" or "The Science of Well-Being" in their titles.[54] Yet, these books often differ in how they define happiness, what methods of measurement they prefer, what they consider the science of happiness to be, and the conclusions they reach. Reason the more, therefore, for readers to look at the new books as presenting different views of *a* science rather than of *the* science of happiness. Viewing them in this way allows us not only to compare them and see what we can learn from them, but also to reflect on more inclusive ways to conceive of such a science. Just as we have much to gain from the many understandings, old and new, of how happiness is experienced and how it should be defined, so it matters to consider the different views of what a science of happiness should include and, in turn, exclude.

Among inclusive views of such a science, the most ambitious ones may well have been formulated in the heady intellectual climate of eighteenth-century Europe.[55] Beccaria, Diderot, Voltaire, and so many Enlightenment figures took for granted that the natural sciences would play a central role, alongside philosophy and the arts, in promoting human happiness; and that it was in the confluence of works of literature, autobiography, politics, philosophy, religion, and science that the subject would be most provocatively and imaginatively explored.

One who called for a new "science of happiness" was himself a theologian, the abbé Pluquet. In 1767, as the newly installed holder of the chair of moral philosophy at the Collège de France, Pluquet published a two-volume book on "sociability," hailing the "revolution of the human spirit" that had become possible because of the union between philosophy, literature, and the methods of the exact sciences.[56] Together, he wrote, those who worked in these disciplines were producing a new ethics, bringing together "the scientific spirit and the love of beauty that will enable a veritable science of happiness to be accessible to all of humanity."

Pluquet's triumphant hopes have not yet been fulfilled; and his aim for a broad collaboration between up-to-date modern disciplines was ahead of its time by over two centuries. If he could return today, he would find the vastly expanded interdisciplinary work among social and natural scientists to be of the highest interest. So would d'Alembert and Diderot and many of the contributors to the great *Encyclopedia* (as its longer title *Encyclopédie, ou dictionnaire raisonné des sciences, des arts, et des métiers* makes clear).[57] They would applaud the current references to a "science" of well-being or happiness, though find it amusing that such a science should seem so new in the twenty-first century. But they might also signal two lacunae in the many recent works on such a science: the failure to make greater use of the rich resources from literature, the arts, history, and philosophy; and the short shrift given to questions about the links between happiness and virtue or moral excellence that were once so forcefully debated.

Might an inclusive contemporary science of happiness bring both sets of concerns back into the debate? Such an approach would make use of all available quantitative findings and span the various disciplinary perspectives to take the full measure of the scope of happiness. It would seek out areas where natural and social scientists and humanists can work together, finding points of convergence and challenges to inadequate or superficial conclusions. Scientists can test the generalizations of humanists when they make assertions of fact; the latter, in turn, can challenge scientists' leaps from statistical averages and correlations to generalizations about human conduct and paths to happiness. The resulting dialogue would reconfirm the centrality of the moral issues inherent in the pursuit of happiness; and draw not only on current research but also on the views of poets, artists, and religious and secular thinkers from the earliest days of reflection about the human condition.

Robert Nozick crossed all such disciplinary boundaries decades before most philosophers paid attention to what social scientists and neuroscientists had to say about happiness, taking their research into account, using it in teaching, while returning again and again to his Experience Machine thought-experiment as a vehicle for working out his ideas about happiness. By now, a number of philosophers writing about happiness are engaging in such interdisciplinary collaboration. Among them, Anthony Appiah brings together works in philosophy, the classics, and humanities more generally, with experimental work in psychology and neurology, in *Experiments in Ethics*. Owen Flanagan suggests, in *The Really Hard Problem*, that Darwin's theory of evolution by natural selection provides prospects for the unification of all the sciences that pertain to human beings. And Daniel Haybron argues, in *The Pursuit of Unhappiness*, for ways in which philosophical reflection and empirical psychological research on happiness can be mutually enriching, in part by questioning conclusions arrived at in the absence of outside challenges.[58]

A century ago, William James offered such mutual enrichment in his lectures on *Varieties of Religious Experience*, asking about the degree to which views of happiness and misery might be influenced, not only by personal experience but also by variations in temperament. Drawing on his vast storehouse of "human documents" old and new, he reflected on the intricate ways in which people's perceptions of happiness are shadowed or sunlit by their own dispositions:

> The sanguine and healthy-minded live habitually on the sunny side of their misery-line, the depressed and melancholy live beyond it, in darkness and apprehension. There are men who seem to have started in life with a bottle or two of champagne inscribed to their credit; whilst others seem to have been born close to the pain-threshold, which the slightest irritants fatally send them over.[59]

VI

BEYOND TEMPERAMENT

The Elusiveness of Black Bile

MODERN SCIENTISTS HAVE WORKED OUT IN CONSIDERABLE DETAIL the hypothesis that William James already called highly likely: that temperament is "organically weighted" on the side of cheer or misery, bringing utterly different perspectives bear on the experience of happiness, how to define it, and whether it is even achievable.[1] He contrasted the "sanguine and healthy-minded" to the "depressed and melancholy." Happiness researchers often distinguish between "extroverts" and "neurotics," finding the former more outgoing, sociable, and likely to experience positive emotions than the latter, who experience negative emotions such as worry, fear, or sadness more frequently.[2] Both clusters of traits are seen as more strongly correlated with differences in subjective well-being than, for example, age, sex, or income; and remain relatively stable over people's lifetimes. But many questions remain about how these traits develop in individuals, how they interact with other personality traits such as creativity, resilience, empathy, and risk-taking, and how they influence people's overall sense of happiness.

The search for a physiological explanation of personality differences began in antiquity. Hippocrates, the celebrated physician who headed a school on the Greek island of Kos (and whose Hippocratic Oath is still invoked for graduating doctors), held that four bodily fluids – blood, black bile, yellow bile, and phlegm – interacted with what was hot, cold, wet and dry to influence the sanguine, the melancholic, the choleric, and the phlegmatic temperaments.[3] Black bile and blood were thought to affect states of unhappiness and happiness most directly.

Robert Burton tells the story, in his 1621 work, *Anatomy of Melancholy*, of how Hippocrates went to visit Democritus, "the laughing philosopher" of Abdera, who taught that cheerfulness was the highest goal in life.[4] Hippocrates was amazed to find Democritus gloomily sitting in his garden under a tree from which hung parts of cats and dogs and other animals that he had been dissecting in the hopes of finding evidence of black bile. If he could locate the seat of *atra bilis* or melancholy and learn how it functioned, he hoped to find a cure for the condition in himself and by his observations teach others how to prevent it and cure it.

Burton, who first published his book under the pseudonym "Democritus Junior," meant to carry on that search for black bile. Alas, the effort was doomed to failure. There was no such bodily fluid. Black bile was a theoretical construct, an invention needed to bring the number of fluids to four to coordinate with the four temperaments or humors and the four influences of hot, cold, dry, and wet. Being anything but an experimentalist, Burton conducted his own search among books, not carcasses. In his thousand-page *Anatomy*, complete with "Partitions," "Members," and "Subsections," he explored the ways in which the four bodily fluids and temperaments mixed and interacted and the strange, sometimes devastating effects of these interactions on human experience. The result was a vast, astounding book, glittering with commentary on every conceivable subject, speculating about different types and degrees of states ranging from the heights of joy to wistfulness, sadness, and what we would now call clinical depression.

Drawing on his own experience, Burton testified to what scientists have since shown to be the case, namely that there is no correspondence between melancholy and sanguine feelings such that more of one automatically produces less of the other. On the contrary, it is possible to experience both at once, as in John Keats "Ode to Melancholy":

Ay, in the very temple of Delight
   Veiled Melancholy has her sovran shrine,
      Though seen of none save him whose strenuous tongue
   Can burst Joy's grape against his palate fine . . .

If Burton could have looked ahead two centuries, he would surely have quoted Keats on this score, along with invoking and embroidering on the testimony of so many others, from Democritus to Montaigne. Like these two, Burton was a "laughing philosopher" who used his study of melancholy to diminish its power over him whenever it struck. At the outset of his book, in "The Author's Abstract of Melancholy," he speaks of knowing both states of mind:

When to myself I act and smile,
With pleasing thoughts the time beguile,
By a brook side or wood so green,
Unheard, unsought for, or unseen,
A thousand pleasures do me bless,
And crown my soul with happiness.
   All my joys besides are folly,
   None so sweet as melancholy.
When I lie, sit, or walk alone,
I sigh, I grieve, making great moan,
In a dark grove, or irksome den,
With discontents and Furies then,
A thousand miseries at once
Mine heavy heart and soul ensconce,
   All my griefs to this are jolly
   None so sour as melancholy.[5]

Philosophers, Genius, and Melancholy

Melancholy has had special appeal for philosophers aware of a question attributed to Aristotle: Why is it that all men of genius, like Empedocles, Plato, and Socrates, who have become outstanding in philosophy, politics, poetry, or the arts, are melancholic, some even to the extent that that they are "infected by the diseases that arise from the black bile"?[6]* By the time of Cicero, a simplified version had become dogma: Aristotle "says that all geniuses are melancholic."[7] Seneca, Burton, and many later thinkers cited Aristotle and indicated that they, too, were of such a melancholy temperament.[8] After all, otherwise they would seem to have lost the chance to be classified among those who have attained greatness. As Burton put it, Aristotle had said that "melancholy men of all others are most witty, which causeth many times a divine ravishment, and a kind of *enthusiasmus*, which stirreth them up to be excellent philosophers, poets, prophets, etc."[9] But genius need not preclude experiencing every shading of happiness as well. The challenge was to remain capable of responding to misery and joy alike without one's awareness of either becoming so overpowering as to dull or obliterate one's response to the other.

Long before publishing his systematic works on ethics, Immanuel Kant examined how that challenge arose for persons of different temperament, in *Observations on the Feelings of the Beautiful and the Sublime*.[10] Unlike Burton and others who drew on Aristotle, Kant never claimed that a melancholy disposition was somehow indispensable for genius. And he made short shrift of the popular notion that persons of such a disposition live steeped in gloom, insensitive to beauty. Instead, he associated melancholy with a profound feeling for the beauty and dignity of human nature and with virtue, while linking good-heartedness and the capacity to be moved to sympathy with the sanguine temperament. The former disposition at its best and most steadfast is sublime; the latter, more changeable according to circumstances, is beautiful.

---

* Aristotle's passage indicated that melancholy could lead some to frenzy and even madness, as when Heracles, in a fit of rage, murdered his wife Megara and their children. Plato himself, in *Phaedrus*, had held that "the greatest of blessings come to us through madness when it is sent as a gift of the gods" and that those without divine madness who hope to be good poets meet with no success: "the poetry of the sane man vanishes into nothingness before that of the inspired madman."

Although Kant spoke of the melancholy disposition with obvious personal identification, he was scrupulous in showing how such a character can deteriorate. For someone who cannot maintain the full sharpness and delicacy of the feeling for the dignity of human beings and live guided by reason, "earnestness inclines toward dejection, devotion toward fanaticism, love of freedom to enthusiasm. Insult and injury kindle vengefulness in him. He is then much to be feared. He defies peril, and disdains death."[11] The person of sanguine temperament can deteriorate as well, but in a different manner: he can slip away gently from friends who are ill or in adversity until the circumstances have changed; if he is wicked, it is "more from complaisance than from inclination."[12]

The nineteenth-century German philosopher Arthur Schopenhauer embodied the deepest melancholy, projecting his dark perspective to encompass all of humankind in near-cosmic gloom. In "On the Vanity and Suffering of Life," he declared that "Everything in life proclaims that earthly happiness is designed to be frustrated, or recognized as an illusion."[13] True, Schopenhauer admitted, in a late essay on the wisdom of life, he had once written that "to secure and promote a feeling of cheerfulness should be the supreme aim of all our endeavors after happiness." Now he knew the "sublime melancholy which leads us to cherish a lively conviction of the worthlessness of everything, of all pleasure and of all mankind, and therefore to long for nothing but to feel that life is a burden that must be borne to an end that cannot be very distant." Like so many others, he pointed, as if for reassurance, to Aristotle, interpreting him as holding "that genius is allied to melancholy and people of very cheerful disposition are only intelligent on the surface."[14] Persons of genius should cherish solitude, avoiding the snares and pitfalls of friendship and above all marriage.

Contemporary research shows Schopenhauer to have been wrong in his exorbitant view that everything in life proclaims that earthly happiness is designed to be frustrated. Like many who postulate unprovable empirical suppositions as self-evident, he included a rejoinder that illusion must be blinding those who disagree. He was equally wrong about the necessary links between melancholy, a life of solitude, and genius. He could have looked to his own contemporary, Charles Darwin, happily married and a devoted father, among any number of non-solitary but remarkably creative individuals – some of whom, like Darwin, surely qualify as geniuses. By now, psychological research shows a clear correlation between higher levels of felt happiness and creativity, as opposed to the near-paralysis that can come from states of depression.[15] A recent study of the advice Schopenhauer offered for practical living concludes that little of it appears to fit the conditions for happiness: following his advice "would probably make us unhappier, even if we had the same neurotic personality."[16]

In his own life, however, Schopenhauer did come as close as any to personifying the Aristotelian linking of melancholy not only with genius but also with the frenzy mistakenly attributed to diseases arising from black bile. Few could rival his eloquence and subtle philosophical insights; but he spent much of his life sunk in unproductive despondency and resentment, and could work himself into states of despairing rage directed against family members, other thinkers, and, always, women. Along with all the other hateful characteristics he attributed to women, he maintained that they were incapable of such happiness as others attained. The keenest joys and sorrows were not, alas, for them, since they had a limited capacity for reasoning and remained "big children all their life."[17] Bertrand Russell, pointing to Schopenhauer's gloomy view that the human will is thoroughly evil and responsible for the suffering that inevitably accompanies life, could not resist adding that as to sex, "this was a wicked business too, because procreation merely provided new victims for suffering. Connected with this view is Schopenhauer's misogyny, for he felt that woman's part in all this was more deliberate than man's."[18]

Melancholy and Depression

In *The Noonday Demon: An Atlas of Depression*, Andrew Solomon explores what is now known about the many shadings of melancholy, from the mildest wistfulness to the severest forms of clinical depression. Drawing on research results from dissection, surgery, and the neurosciences, he uses the geographical metaphor of an atlas rather than Burton's anatomical model. Like Burton and Democritus, Solomon studied his own condition for a double purpose: though writing on depression was painful, sad, lonely, and stressful, the idea that he was doing something that might be useful to others was uplifting even as it was useful to himself as well. At one point, an experience of happiness was so unfamiliar that he did not recognize it as such: "Years had passed since I felt happiness at all, and I had forgotten what it is like to live, to enjoy the day you are in and so long for the next one, to know that you are one of the lucky people for whom life is the living of it."[19]

Neuroscientists examine states of depression and elation as reflected in activities in different regions of the brain. They show, for example, how certain areas of the brain light up during pleasurable experiences or memories of such experiences, but not during states of "anhedonia," in which subjects find nothing either pleasurable, achingly beautiful, delicious, or joyous. By measuring the rate of blood flow in different parts of the brain, they compare the quite different responses of depressed and other persons to tasks such as reading a list of somber, depressing words; and chart the debilitating long-term effects on the brains of war veterans suffering from post-traumatic stress disorder. This condition was first classified as a psychiatric disorder among returning Vietnam War veterans, some of whom found it impossible to return to ordinary life, impossible to engage in close human relationships, impossible to deal with the memories of the violence that they had experienced, impossible to stop reliving those memories. By now, the condition is seen as affecting a number of victims of rape and assault more generally, as well as of children who witness violence in their families and their neighborhoods.[20]

The evidence brought to light by neuroscientists resonates with the ancient views of the role of blood flow in making for differences between people of melancholy and sanguine temperaments. The metaphor attributed to the poet Alice Meynell – "Happiness is

not a matter of events; it depends upon the tides of the mind" –
likewise evokes the ebb and flow of the dopamine and the serotonin systems in the brain, and their influence (along with that of
cortisol and other stress hormones) on the immune system and
on mood, with over a hundred brain chemicals and countless
neurotransmitters communicating across trillions of synapses in
the brain. These fluctuations can be shown through neurological imaging, as can the experience described by Sylvia Plath, of
feeling as if her life "were magically run by two electric currents:
joyous positive and despairing negative – whichever is running
at the moment dominates my life, floods it. I am now flooded
with despair, almost hysteria, as if I were smothering. As if a great
muscular owl were sitting on my chest, its talons clenching &
constricting my heart."[21]

In the half-century since Plath's suicide at thirty-one, there
have been great advances in the treatment of depression in the
fields of psychopharmacology, and cognitive therapy, as well as
improved methods of electroshock therapy for the most severely
afflicted individuals. Pharmaceuticals that might relieve or elimi-
nate the debilitating memories that go with post-traumatic stress
are also being investigated.[22] The more we learn about the differences between mild melancholy and clinical depression, the
starker the discrepancy becomes between what is being done and
what should be done to bring help to those most in need –
between those who are being successfully treated for depression
and the vast numbers the world over who have no access to treatment. Major depression is now the leading cause of disability
globally.[23] Projections for the coming twenty years indicate that it
will become second only to heart disease among causes of death
and disability. Depression and schizophrenia are responsible for
60 percent of all suicides. In poor countries, depression is often a
cause of shame, which prevents individuals from seeking such
help as might be available. And even in the richest societies, it is
estimated that for every six persons suffering from clinical depression, only one receives the most appropriate treatment, whereas
two are treated inappropriately, and three go untreated.

Sanguine Dispositions

I am no Plant that will not prosper out of a Garden. All places, all airs, make unto me one Countrey; I am in England every where, and under any Meridian. I have been shipwrackt, yet am not enemy with the Sea or Winds; I can study, play, or sleepe in a Tempest. . . . In brief, I am content, and what should Providence add more? Surely this is what we call Happiness, and this do I enjoy.

Sir Thomas Browne, *Religio Medici*, 1643[24]

Sir Thomas Browne was a practicing physician and man of genius who, *pace* the Aristotelian doctrine, was anything but melancholy. He had an exuberant interest in everything and everyone in the world around him. While his somewhat older compatriot Burton had conveyed intimate knowledge of both the melancholy and the sanguine temperaments, Sir Thomas spoke in glowing terms of how differently he experienced reality from most others. His exultant claims to withstand tempests and every hardship may test our credulity, but his effervescence, cheer, and vivid imagination are bound to strike his readers. In *Religio Medici*, he aimed to sort out what he could and could not believe about religion and to understand more about the amazing world of nature around him and its inhabitants, including his fellow humans and himself. He appears to have been unusually able to enjoy life and to find happiness, while steering clear of hatred, envy, dejection, and despair.

One reason Sir Thomas gave for his felicity, in spite of the suffering he saw as a doctor and the wars and persecutions at home as abroad, was his religious faith. It assured him, he wrote, that he would go to Heaven rather than to Hell after death. He felt able to fall asleep each night, trusting that he would either wake once more to his own happy life or wake forever to greater happiness still, in the presence of God. But more must have been at issue than religious faith, or else his fellow Christians would all exhibit the same sanguine disposition. Not for him any brooding obsessions with sin, perdition, and hellfire, much less the conviction among Calvinists and Puritans that they might have been predestined for the tortures of Hell. Nor did he have the slightest inclination to persecute those not sharing his faith. He felt no enmity to anyone on grounds of religion or nationality, claiming to have an equal toleration for all theological systems.

Even such generosity of spirit, however, would not by itself explain the zest for living he so clearly enjoyed and the sense of happiness he described. His rich imagination and humor surely also contributed to his felicitous disposition, making it possible for him to shift perspectives, to look at the world and at himself in different lights. Humor, especially, is correlated with subjective well-being, along with self-esteem, creativity, and a tolerance for ambiguity, diversity, and change – precisely the traits with which Browne was so amply endowed.[25]

Commenting two centuries later on Sir Thomas's claims to surpassing felicity, Sir Leslie Stephen, editor of the 1882–1900 *Dictionary of National of Biography* and himself no stranger to periods of depression, suggested that no matter what skepticism these claims might evoke, they deserved to be taken seriously:

> For if we set aside external circumstances of life, what qualities offer a more certain guarantee of happiness than those of which he is an almost typical example? A mind endowed with an insatiable curiosity as to all things knowable and unknowable; an imagination which tinges with poetical hues the vast accumulation of inco-herent facts thus stored in a capacious memory; and a strangely vivid humour that is always detecting the quaintest analogies; and, as it were, striking light from the most unexpected collocations of uncompromising materials: such talents are by themselves enough to provide a man with work for life, and to make all his work delightful.[26]

The French Jesuit priest and world-renowned scientist Pierre Teilhard de Chardin exhibited just that insatiable curiosity, that imagination, capacious memory, and vivid humor. He traveled the world, taking part in paleontological excavations including those that led to the finding of Peking Man, lecturing and writing on science and on the evolution of human beings and of the cosmos, dealing as best he could with Church officials who forbade him to publish his writings during his lifetime. In so doing, he reflected, like Browne, on the nature of happiness and on how temperament and experience influence how it is perceived.* In a 1943 lecture, "On Happiness," Teilhard pointed to the ancient phrase "De vita beata" (On the happy life), asking what in fact would make for such a life.[27] He posited three attitudes to life and happiness for which we all have the germs within ourselves: the tired, constricted view, at its extreme resembling Schopenhauer's pessimism but more often simply seeking the tranquility that comes from cutting back on needs, feelings, and desires; the approach of the hedonist or pleasure-seeker, enjoying each moment and each thing for their own sake; and the enthusiast's conception of living as an ascent and a discovery, for whom the happy man is one "who, without any direct search for happiness, inevitably finds joy as an added bonus in the act of forging ahead and attaining the fullness and finality of his own self."[28]

Members of this third group, Teilhard suggested, share his own zest for life in all its complexity – zest he defined elsewhere as "that spiritual disposition, at once intellectual and effective, in virtue of which life, the world, and action seem to us, on the whole, luminous – interesting – appetizing."[29] The feeling and development of such zest are to some extent, he believed, our own responsibility. We need to cultivate ourselves, bringing more order and unity into our ideas, feelings, and behavior, become capable of uniting with others and of love in all its forms, and attach our lives to something larger than ourselves. Among the individuals whom Teilhard mentioned as partaking of the happiness that zest brings were explorers in the Arctic and the Antarctic such as Nansen and Shackleton, aviator pioneers who conquered the air, and scientists,

---

* And just as Browne was convinced that he would find happiness, not punishment, after death, so Teilhard had the confidence to say, shortly before his death, that "If in my life I haven't been wrong, I beg God to allow me to die on Easter Sunday." He died on Easter Sunday, April 10, 1955.

such as "the two Curies – the husband and wife who found happiness in embarking on a venture, the discovery of radium, in which
they realized that to lose their life was to gain it."[30] To be happy,
Teilhard held, we must seek to add one stitch, no matter how small
it be, to the magnificent tapestry of life.

Even persons of such exceptional zest and enjoyment of their
life's work as Browne and Teilhard knew times of disappointment
or despair. Neurophysiologist Felicia Huppert, seeing positive
mental health as combining subjective well-being and realizing or
developing one's potential, points out that zest and other positive
feelings are not by themselves sufficient for this purpose. Such feelings may be transitory, as when induced by drugs, and there are
occasions when positive mental health requires the experience of
negative emotional states such as grief.[31] The ability to experience
other negative emotions, such as fear, is likewise indispensable not
only for mental health but for alertness to every sort of danger.

Individuals vary greatly in how self-protectively they shield
themselves from fear and other negative emotions. Persons such as
Browne and Teilhard possess to a high degree what is often called
resilience, the ability to bounce back from adversity of all kinds,
including the severest trauma. In a passage that he suppressed from
his *Religio Medici*, thinking perhaps that it might sound too
boastful, Browne claimed that "I have that in me that can convert
poverty into riches, adversity into prosperity, and I am more invulnerable than Achilles; fortune hath not one place to hit me."[32]

A strong measure of resilience is surely necessary for sheer
survival. When it is impaired, as in post-traumatic stress disorder,
there is no end to the reliving and re-envisioning of the traumatic
events. The shock of trauma can disable switching mechanisms in
the brain that ordinarily soften painful recollections – as is well
known, for instance, about our failures to remember accidents,
operations, even labor pains. If these brain mechanisms malfunction, individuals suffer from persistent searing memories and
anguish.[33]

Although psychological resilience serves a crucial survival function, it has a dark side: it can operate so efficiently as to blot out

awareness of anything that might trouble one's conscience, any remorse about past deeds, qualms about immoral aims for the future, or concern for the suffering of others more generally. Those who share Browne's thorough capacity to enjoy life, his sublime self-confidence, and his reliance on what he called "oblivion" to erase life's disagreeable or tragic aspects may indeed lead happier lives than anyone lacking such strong defenses.[34] Yet their self-shielding may also blind them to human realities that would otherwise stare them in the face. For example, Browne's high-powered resilience may have lessened his compunction about using his medical expertise to certify that two women suspected of witchcraft were in fact witches and should be convicted, even though he rejected the "proofs" adduced to support the belief in witchcraft and must have known that torture and death by hanging, strangling, or burning would be these women's fate.

Though oblivion, compartmentalization, and other forms of denial are hardly unique to people of a sanguine disposition, their high level of resilience makes it easier to give free play to the temptation to "see no evil, hear no evil." Even as resilience can make people better able to deal with adversity and unhappiness in their own lives, it can render them inattentive to the needs of others. It is then dangerous in the extreme. William James pointed out that optimism may become quasi-pathological in some individuals: "The capacity for even a transient sadness or momentary humility seems cut off from them as by a kind of congenital anaesthesia. . . . [H]appiness, like every other emotional state, has blindness and insensibility to opposing facts given it as its instinctive weapon for self-protection against disturbance."[35]

Balancing Resilience and Empathy

To avoid the self-protective drive to shut out awareness of the needs of others, resilience must be counterbalanced by empathy, the capacity for fellow-feeling and compassion. That capacity lies at the root of morality and of the moral precepts of Buddhism, Christianity, and many religions. In the fourth century BC, the

Chinese philosopher Mencius spoke of compassion as one of the four "hearts" or feelings without which people are not fully human, along with shame, modesty, and a sense of right and wrong. If we deny these potentialities instead of developing them, he warned, we cripple ourselves.[36] Immanuel Kant likewise held that moral feeling and the love of one's neighbor were among the natural predispositions that make it possible for us to be affected by concepts of duty – necessary for morality but not sufficient for its exercise.[37]

Both resilience and empathy are present in rudimentary form at the neurological level even before birth. We know the effects on the brain when they are nurtured, strengthened, or on the contrary neglected, stunted, even completely eroded through misfortune and abuse.[38] Individuals found to have higher levels of subjective well-being are often more aware of other individuals, more empathetic, at least with respect to those individuals whose needs come to their attention. But at all levels of subjective well-being, empathy can be felt toward some at the cost of others, as when partners of abusive parents cannot bring themselves to interfere with the abuse inflicted on their children; or it can become engulfing, especially when it focuses primarily on suffering, rather than on all of human experience, from the most joyful to the saddest. Without some capacity to resist being overwhelmed by compassion for human suffering, people can become distraught to the point of being of little help either to themselves or others. In his novel *Time Must Have a Stop*, Aldous Huxley describes the reaction of a young man overcome by thoughts of human misery:

> All over the world, millions of men and women lying in pain; millions dying, at this very moment; millions more grieving over them, their faces distorted like that poor old hag's, the tears running down their cheeks. And millions starving, millions frightened, and sick, and anxious. Millions being cursed and kicked and beaten by other brutal millions. . . . The horror was always there, even when one happened to be feeling well and happy – always there, just round the corner and behind almost every door.[39]

The biblical injunction to those who go out into the world to be "wary as serpents and innocent as doves" speaks to the need for both the self-protection that resilience affords and the concern not to do harm to others that is rooted in empathy. To be wary and innocent, or as other translators of the Bible have put it, cunning and guileless: this seems to be paradoxical advice. Ordinarily, we think of the wary and the innocent as quite separate types of persons, each bound for pitfalls, either of excessive promotion of their own interest or of naïve vulnerability. But if we try to unite these two qualities, and to strive for them jointly, then we see that together they delineate a state of integrity or wholeness – the capacity to be wary enough to avoid being harmed, and innocent enough not to inflict harm. Having either a sanguine or a melancholy disposition without such balance can skew perspective and vitiate moral choice. In Voltaire's *Candide: or, The Optimist*, the ever-cheery Pangloss is as dangerous as the resigned Martin, expecting nothing but the worst from what he sees as vain human struggles to seek happiness.

Yet more than such balance between resilience and empathy is needed in reflecting on practical moral questions about happiness. How should I weigh my own happiness against that of others? How, beyond their happiness, do their rights and needs enter into such weighings? At what cost to others is it permissible to pursue my own happiness or that of my family or community – say when it comes to energy usage and the environment? What if I could indeed feel happier to the extent that I learned to steel myself against perceiving the needs of others? Or against awareness of how I might have hurt them? Many authors offering advice on happiness leave such questions entirely out of account. Some who counsel readers to cultivate positive thinking and to see the bright side of things characterize people troubled about moral conflicts and personal responsibilities as "neurotic" or as "ruminators." Other authors look primarily at the advantages of adaptive mechanisms and resilience, as when one psychiatrist suggests that they "soften conflicts of conscience" such as that of putting a parent in a nursing home.[40]

To see how awareness of one's responsibility toward others might conflict with adaptive mechanisms that enhance personal happiness, consider an everyday example mentioned by Kant. He asked what he should do after having insulted someone in the company of others, then returned home feeling troubled and wishing for an opportunity to repair the situation. Should he just stop worrying about this incident, set it aside, by invoking the human frailty and infirmity that is common to us all? No, he concluded. One should only take such frailty and infirmity into account when judging the actions of others; when it comes to oneself, one should have recourse to no such excuses, but instead think through where one's duty lies and strive to act accordingly.[41]

Asking oneself what to do when one is uncertain of one's moral duty requires countering the quick intuitive recourse to self-protective strategies such as seeing the silver lining of a problematic action, imagining that it has gone unnoticed or at any rate harmed no one. It calls, instead, for mobilizing efforts to perceive the problem, to think it over or deliberate about it, to arrive at a choice about how best to respond to it, then carry it out. At each step, the temptation is strong to opt for premature closure, bypassing efforts at the fullest perception and the fairest, most careful deliberation, then settling for uncritical acceptance of simplistic choices and inadequate responses.

The same moral considerations hold when it comes to other personality traits, ones that differ from people's tendency to live either on what William James called the sunny side of their misery line or in gloom and desperation. These other traits are also aspects of temperament or disposition and contribute powerfully to how differently individuals conceive of happiness. Among them are differences in how people experience time, with some people living primarily in the present and others focused, at times obsessively, on the past or on the future. Psychologists Philip Zimbardo and John Boyd point to research showing how people's attitudes toward time, their internal time zone, influence everything they do, usually unbeknownst to themselves; the authors suggest that by becoming aware of our own attitudes toward the past, the present, and the future, we can learn how to transform negative experiences into positive ones and how to capitalize on positives in the present and the future without succumbing to blind devotion to either.[42]

Another set of differences has to do with people's attitudes when it comes to order and cleanliness. In daily life, such differences, concerning, for example, tolerance for messy surroundings, chaotic lifestyles and personal hygiene, are reputedly among the chief causes of marital discord and of resentment among college room-mates. Even apart from the minutiae of everyday life, incomprehension has long been the rule between those who would find life meaningless unless they could see it as part of an orderly universe with a purpose, a guiding providence, and others who express full comfort with a less ordered cosmos.

Still other differences involve attitudes toward risk-taking, some of which may be linked to genes that regulate the brain's serotonin and dopamine levels, which influence levels of anxiety and fearlessness. Those rare persons who dream of reaching the summit of Mount Everest would submit to almost any sacrifice or personal hardship for the sake of the delight they anticipate once they succeed. For most others, sacrificing ordinary sources of happiness, possibly life itself, vastly outweighs the chance of experiencing any such thrill; and some people, with high levels of risk aversion, take pains to avoid even the minutest risks.

On each of these scores, proponents of the opposing views often maintain that their very happiness is at stake, regarding those at the opposite end with bewilderment. In turn, the desire for freedom cuts across the preferences expressed by people who differ in all such ways. Robert Mauzi, in setting forth the passionate debates about many such differences in eighteenth-century France, explores with subtlety a related distinction rarely taken up in contemporary studies of happiness: what he calls the duality between movement and rest. He suggests that the most general definition of happiness that would convey "l'esprit du siècle" would be that "happiness is the state of a soul having resolved the fundamental antagonism between the temptation to vertigo and the dream of rest, between movement and immobility."[43]

Often linked to sanguine and melancholy leanings, and in turn to extroversion and introversion, are the preferences people express for companionship or solitude. Psychologists have devoted special attention to the powerful effect of social contacts on happiness. At any point in time, roughly 20 percent of individuals find that social isolation is a major cause of unhappiness, according to the authors of *Loneliness: Human Nature and the Need for Social Connection*.[44] They compare loneliness, or social pain, to physical pain.

More generally, extroverts are taken to be more likely to be happy than introverts. While recognizing the importance of individual differences, researchers stress that, on average, individuals thrive on having adequate social contacts and suffer when isolated. Advice books, accordingly, suggest that if people want to have better chances at happiness, they should go out, make friends, join organizations, take part in community and religious life. Some psychologists even regard social contacts as the one factor without which happiness is not possible. Ed Diener and Robert Biswas-Diener claim, for example, that research results are clear on the point that "healthy social contact is essential for happiness."[45]

What can we learn from viewing these findings against the background of the long-standing controversies about whether it is more conducive to happiness to seek out an active, engaged existence or a life of quiet repose and contemplation? Or the differences among family members with entirely different preferences when it comes to gregariousness or the longing for solitude?

Retirement, Leisure, and Solitude

The Greek word *idiotes* originally referred to persons who stayed away from the public arena and failed to defend the city-state. For Aristotle, such individuals did not begin to live up to the functions of human beings, whom he famously defined as social animals. Nor could anyone be happy who sought a passive existence, since happiness requires the soul's activity and actions that express virtue. But commentators have found his stress on activity difficult to square with his shift, in the last book of his *Nicomachean Ethics*, to holding that the happiest life of all was one of *theoria* – of philosophical contemplation. He compared such an existence to that of the gods, one that was clearly not an option for ordinary citizens and non-philosophers.

By 300 BC, when Greek city-states were in disrepair, the advice of Epicurus to take leave of civic life resonated with many. He invited his followers – women as well as men, slaves as well as free persons – to join him instead, finding happiness through "living retired" in the garden that he had created near Athens. They would live as a community of friends sharing everyday tasks and simple pleasures and thus not in any way be solitary; yet they would take no part in public life.

Seneca, who had endorsed the Stoics' stress on an active life of public service as opposed to the contemplative life, wrote increasingly of the joys of retirement during the last few years of his life. Quoting Epicurus again and again, he held that the time could come when the corruption of the state or one's own inability to serve rendered retirement preferable. And Augustine, while discussing "the happy life" with relatives and friends at Cassiciacum, had hoped to live quietly with a group of such friends, devoting himself to communal life and to writing, only to find himself immersed in active service after he was appointed Bishop of Hippo. Commenting on the biblical account of Mary and Martha, he underscored the choices, sometimes the tensions, between these two ways of life and the Church's need for service of both kinds.

Few have written more feelingly than Petrarch in favor of the kind of life that brings "the blessedness of solitude" as against a life that renders people "agitated and careworn and breathless."[46] He had grown up in the hectic atmosphere of the fourteenth-century papal court in Avignon, in Provence, where his father served as a minor official. When he was thirty-three years old, and already renowned as a poet, he bought a home in the verdant countryside in nearby Vaucluse. His *De vita solitaria* resonates with the joy of being free to devote himself to a contemplative, creative existence – not at all, he specified, some "pining and anxious seclusion" but rather a serene existence in the company of loving friends: "No solitude is so profound, no house so small, no door so narrow but it may open to a friend."[47]

In this chosen environment, Petrarch cherished the leisure to read and write, "alternately finding employment and relief in each, to read what our forerunners have written and to write what later generations may wish to read." Such reading and writing was a form of companionship in its own right, with past thinkers and future readers; as valuable, Petrarch insisted, as serving the civic and political functions that Cicero and others had held forth as patriotic duties. To Cicero's haughty claim that "the abler and more accomplished a man is, the less he would care to be alive at all if debarred from taking part in public affairs," Petrarch responded that one can contribute as much through writing and other work carried out in private life; indeed, it is not only better to live in retirement but a duty for those able and willing to do so, whether for religious reasons or for their own sake and for their serious studies; better to withdraw as far as one can from the haunts of men and crowded cities. But far more drew him to such a life:

> To live according to your pleasure, to go where you will, to stay where you will; in the spring to repose amid purple beds of flowers, in the autumn amid heaps of fallen leaves; to cheat the winter by basking in the sun and the summer by taking refuge in cool shades, and to feel the force of neither unless it is your choice![48]

In sections on Epicurus, Seneca, Cicero, Augustine, St Jerome, and a great many others, Petrarch addressed the debates among classical philosophers about the relative merits of the active life as against the contemplative life, and, among Christians, of monastic seclusion as against taking full part in public life. While he wrote respectfully of religious contemplatives and often visited the monastery where his brother lived, his own contemplative bent went beyond religious meditation to include, also, the beauties of nature and poetic and philosophical reflection. Not for him, ecclesiastic though he was, any vows of chastity (he had two children who remained close to him throughout his life), nor of poverty or, least of all, of obedience. His *De vita solitaria* is a veritable manifesto for freedom of choice – the choice of where, how, and with whom to engage in the pursuit of happiness.

Eloquent about his desire to be free to choose how to lead his life, Petrarch was among the relatively few who could hope for such freedom. Most individuals who ended up either in the active or the contemplative life had no such choice, least of all women in his time. Even when they had an opportunity to exercise choice of any kind, the only two paths open to many young girls were marriage and entering a convent. It took women of exceptional strength to find a path of their own, as did St Catherine of Siena, born in 1347, a year after Petrarch's book was published. With her twin sister Giovanna, the youngest in a family with twenty-two children, she refused both marriage and convent life. Instead she asked to be allowed to live alone, as could widows, and to be free to devote herself to her religious life as a mystic, combined with that of caring for the destitute and sick, and, increasingly, that of a forceful advocate of Church reform.

Art and Exhilarated Contemplation

Oh! It used to make me mad with joy when I saw the clouds come up in the sky one by one. I felt, even in those early days, that I was surrounded with a companionship very intense and very intimate, though I did not know how to name it. I had such an exceeding love for nature that I cannot think in what way to describe it to you; but she was a kind of loving companion, always with me, and always revealing to me some fresh beauty.

Rabindranath Tagore, *Letter*, 1929[49]

Regardless of whether people opt for solitude or company, and quite apart from any desire to engage in contemplation as religious meditation or philosophical reflection, many have sought happiness through contemplation in a different sense – that of seeing, observing, attending to visual experience. Dictionaries define contemplation as either intransitive, as in the case of meditation, or as transitive, where it is directed at an object of some kind, as in contemplating *something* – a sunset, a face, a scene of horror or of beauty. It is this transitive sense of the term that Peter Quennell intends, in *The Pursuit of Happiness*, citing the advice of the painter Walter Sickert to his young disciple Nina Hamnett: she should adopt his attitude of "coolness and leisurely exhilarated contemplation."[50]

Such intense, focused contemplation can play a central part in religious meditation, as when Buddhist monks practice visualizing paintings or sculptures of the Buddha, or in the many efforts to imagine and convey conceptions of the Garden of Eden. Taken by itself, however, the active, transitive, exhilarated contemplation which Quennell identifies as that of the artist and the aesthete can relate in its own way to happiness. "One of the noblest gifts that have ever contributed to happiness is the gift of seeing, and sometimes recording, the beauty and wonderful variety of the universe into which we have been born."[51]

Georgia O'Keeffe conveys her experience of that gift in her art and in speaking of what she saw – as in describing climbing a hill in the evening light and coming upon the "sun under the clouds – the color effect was very strange – standing high on a pale green hill where I could look all around at the red, yellow, purple formations – miles around – the colors all intensified by the pale gray green I was standing on." She had no need for the comfort of the Church, she wrote to a friend: "When I stand alone with the earth and sky a feeling of something in me going off in every direction into the unknown of infinity means more to me than any organized religion gives me."[52]

Winslow Homer wrote to his brother at fifty-nine, having decided to settle, and paint, on the rocky coast of Prout's Neck, Maine: "The life I have chosen gives me my full hours of enjoyment for the balance of my life – the Sun will not rise, or set, without my notice, and thanks."[53] When I came across that letter, written in 1895, at the Clark Museum in Williamstown, Massachusetts, it was placed on a wall next to his extraordinary seascape *Sunset, Saco Bay,* painted the following year. I was struck by those words "notice, and thanks." How I would have wanted to see him in the act of giving such "notice and thanks" to sunrises and sunsets and all that he saw.

Peter Quennell averred, in his memoir, that he was an "idiotes" in the Greek sense of a person in a private station; and that he had come to consider art as the only form of endeavor worth pursuing. Since childhood, he had been in love with pictures, "from the great visions that create a whole new world of harmony, order, and celestial peace to family portraits, conversation groups, and the type of landscapes that German art historians call *Wunderlandschaft* where the onlooker is invited to penetrate the background of a strange romantic landscape and literally 'wander round' its details" as in works by Poussin such as that of the ashes of Phocion being removed from Athens.[54]

Images remained for Quennell a major source of happiness, requiring the same effort at contemplation for the viewer as for the artist – one that resonates with Bernard Berenson's resolve, as a student, to follow Walter Pater's injunction to fill his life with the contemplation of beauty and exquisite sensation and to "burn always with the hard gem-like flame, to maintain this ecstasy."[55] Berenson did maintain that ecstasy throughout his long life, as his diaries show. At eighty-four years of age, for example, he wrote of his pure joy while revisiting cities of northern Italy: "Venice was pure ecstasy. I felt as if I had painted it all myself. I lived, enchanted. Not a nook or cranny that it did not make me happy to look at, and to caress with my eyes."[56] As Iris Origo noted, in describing her excursions with Berenson as a child, he conveyed the same intense, focused attention to works of art, landscapes, and everything that he saw that he had had as a student, teaching by example what he meant by the art of looking.[57]

That such intense aesthetic experiences can contribute to the experience of happiness for some individuals should go without question. Philosopher Alexander Nehamas, in *Only a Promise of*

Happiness: The Place of Beauty in a World of Art, speaks of the necessity, for his own happiness, of making beautiful things part of his
life, quite apart from other values, including ethical ones. He ends
his book as follows: "For Socrates, virtue was nothing but its own
pursuit. And only the promise of happiness is happiness itself."[58]

Philosopher Wayne Sumner argues that the lives of people who
possess aesthetic sensibility should also count as having value in
the eyes of outsiders. In *Welfare, Happiness and Ethics,* he distinguishes four dimensions of value that lives can have: prudential
value, or how well a life is going for the individual whose life it is;
aesthetic value; perfectionist value, when lives exemplify excellence of different kinds; and ethical value.[59] The fact that a life can
have high value in any one of the four dimensions entails nothing
about its score in the other three, he persuasively maintains,
however often philosophers have dreamed of showing that the
first and the fourth coincide. Pointing to the scope and variety of
possible preferences, he suggests that many might best be left
unspecified in theories of welfare, whether they be for "planning
over spontaneity, for complexity over simplicity, for civilization
over tribal life, for excitement over tranquillity, for risk over
safety, for perpetual striving over contentment, for sexuality over
celibacy, for companionship over solitude, for religious conviction over atheism, for rationality over emotion, for the intellectual
life over the physical, or whatever."[60]

Such differences are as relevant to the study of happiness
today as in the past. They help explain the incomprehension that
often reigns among individuals with sharply different inclinations
toward, say, risk-taking, orderliness, civic engagement, contemplation, or beauty. How do these inclinations relate to happiness,
for better or worse? How do they interact in individual lives
and affect the lives of others? The richness of personal testimony
and of literary portrayals of how people experience these traits
challenges every quick generalization.

When we explore such accounts, alongside the analysis and
efforts to define and sort out what underlies them by philosophers
and psychologists and others, it becomes possible to strive for

both broader and deeper understanding of how they relate to happiness. In this process, scientists can help set aside claims once made about temperament and happiness or unhappiness as simplistic or unfounded, as is the case with the assumption about the linkage between genius and melancholy attributed to Aristotle and repeated by Schopenhauer and others. Likewise, the insights from philosophy and literature demonstrate the need for caution in the face of advice that appears to encapsulate the potential for happiness in particular traits such as optimism or gregariousness.

For persons considering their own choices, however, it matters to seek to understand how such dispositions affect their lives and those of others. To what extent might, for example, debilitating forms of dejection or risk-taking be inherited or otherwise impossible to overcome? Is it in fact possible for individuals to bring about lasting increases in their own levels of happiness? And might it be possible for all or most people to do so if they follow certain steps or teachings? These questions have been debated since the beginnings of reflection on human happiness; they are ones about which philosophers, theologians, and social and natural scientists still disagree today.

VII

IS LASTING HAPPINESS
ACHIEVABLE?

Freud versus Russell

One feels inclined to say that the intention that human beings
should be happy was not included in the plans of the "Creation."
Sigmund Freud, *Civilization and its Discontents* (1930)[1]

It is in the belief that many people who are unhappy could become
happy by well-directed effort that I have written this book.
Bertrand Russell, *The Conquest of Happiness* (1930)[2]

IN BOOKS PUBLISHED THE SAME YEAR, SIGMUND FREUD AND
Bertrand Russell drew on personal observation and at times flam-
boyant speculation to offer diametrically opposed conclusions
about human chances for happiness. The two works appeal to
different audiences and are rarely if ever read together. But it is
worth revisiting them together, with a special focus on their claims
about happiness. To what degree might their authors have been
influenced by their own personal circumstances and by their
personalities? And how might we look at their conclusions in the
light of current thinking on happiness?

Both men were quick to belittle the importance of their books.
Freud immediately described his "enquiry concerning happiness" to
a friend as seeming to him quite superfluous compared to his earlier
books. He had written this essay while on vacation without access to
a library, about "civilization [Kultur], guilt feelings, happiness and
similar lofty things . . ."[3] And Russell pointed out in his Preface that

"No profound philosophy or deep erudition will be found in the following pages. I have aimed only at putting together some remarks which are inspired by what I hope is common sense."[4]

Over the ensuing decades, philosophers have taken Russell at his word, dismissing his book as nothing but pop psychology. By contrast, Freud's text has achieved iconic status as a twentieth-century classic, not so much for his analysis of happiness as for his views about civilization and the tensions between the imperatives of Eros and Thanatos, between life and love as against the death-drive. According to Louis Menand, the grounds may have eroded for the authority *Civilization and its Discontents* once enjoyed as an ultimate account of how things are, but we can no longer understand the way things are without taking it into account: "Like the Commandatore in Mozart's *Don Giovanni*, its afterlife is in certain respects more impressive than its life, and for the same reason. It can no longer be killed."[5] The same turns out to be true for *The Conquest of Happiness*. Both books continue to enjoy an enthusiastic readership, regardless of the lethal blows that critics have aimed at each one.

The clash between Freud's and Russell's conclusions about human chances for happiness is the more striking as the two started from several shared premises. Both were atheists, convinced that religious belief about divine intentions for humankind and life after death rested on illusion; any discussion of lasting happiness, therefore, had to concern earthly life alone. Both were fascinated by science and would have taken intense interest in the findings of today's happiness research. Both were vivid, imaginative, subtle writers, each nominated for the Nobel Prize in literature: Russell in 1950, when he was awarded the prize; Freud in 1936, after being repeatedly nominated unsuccessfully for a prize in medicine since 1915.[6] Both, finally, took for granted that the great majority of people were unhappy. Russell's claim that "men are so unhappy that mutual extermination seems to them less dreadful than the continued endurance of the light of day" conveys, at first sight, the same air of grim certainty as Freud's assertion about "man's natural aggressive instinct, the hostility of each against all and of all against each."[7]

But that word "natural" in Freud's sentence points to the central difference between their views. For when it came to the possibility of lasting individual happiness, the two thinkers parted company. Whereas Freud concluded that the aggressive instinct inherent in human nature renders all hopes for such happiness fruitless, Russell insisted that this need not be so: it lies within the power of individuals, given average good fortune, to achieve, or "conquer," happiness.

Their disagreement began with incompatible views on how to define or delimit happiness. Freud held a narrowly hedonic view: that pleasure alone could bring spurts, and only spurts, of happiness. "What we call happiness in the strictest sense of the word, comes from the (preferably sudden) satisfaction of needs which have been dammed up to a high degree, and it is from its nature only possible as an episodic phenomenon."[8] More prolonged happiness, Freud explained, is bound to be overwhelmed by suffering that threatens us all from three directions: "from our own body, which is doomed to decay and dissolution . . .; from the external world, which may rage against us with overwhelming and merciless forces of destruction; and finally from our relations to other men."[9] All hopes for long-term happiness are therefore based on illusion, and bound to be continually disappointed. To be sure, human beings are driven to strive after happiness; but only self-blinding and wish-fulfillment allow them to imagine it is attainable.

Russell, like all who share his more complex, eudaimonic view of happiness, agreed that pleasure contributes to the experience of happiness, but argued that it cannot be all that should count. Much more goes into shaping the lasting happiness that he saw as fully achievable: "All our separate tastes and desires have to fit into a general framework of life. If they are to be a source of happiness, they must be compatible with health, with the affection of those whom we love, and with the respect of the society in which we live."[10] The most basic form of happiness is available to human beings, he suggested, regardless of their level of education, so long as they enjoy health and physical vigor, loving relationships, and a sense of achievement in life, of overcoming obstacles. Among the highly educated, different types of activities are needed to generate such pleasures. Happiest, he speculated, was the man of science, who can use his abilities to the full and achieve results that benefit the public at large, not just himself.

The distinction drawn by V. J. McGill – between thinkers who, like Aristotle, have expansive views about happiness and those who insist, with most Stoics, that happiness calls for having the fewest possible needs – bears on the difference between the two men. Russell comes down squarely on the side of the expansive theories and of self-realization. He would have rejected Freud's fatalistic view of the plan of "Creation" as somehow not including the intention that human beings should be happy. Russell might even have labeled such a belief in foreordained misery as religious in its own right, presupposing some power capable of guiding human fates, despite Freud's dismissal of religious belief as mass delusion. Instead, Russell held high hopes for the conquest of happiness, seeing the happy man as one who

> feels himself a citizen of the universe, enjoying freely the spectacle that it offers and the joys that it affords, untroubled by the thought of death because he feels himself not entirely separate from those who will come after him. It is in such profound instinctive union with the stream of life that the greatest joy is to be found.[11]

Such were precisely the thoughts, however, that Freud saw as reflecting an "oceanic feeling," surviving from an infantile sense of unity with the world. Wasn't Russell's notion that it is in "profound instinctive union with the stream of life that the greatest joy is to be found" itself quasi-religious?* And was not the claim that individuals can achieve personal happiness, even in the midst of social

---

* "Exactly," Teilhard de Chardin might have responded. He had spoken with appreciation of Russell's book in *On Happiness*, citing Russell's view that the greatest of all joys is to be found in "a deep and instinctive union with the whole current of life." Teilhard added that Russell, as a "materialist thinker" did not recognize that the union with something larger than oneself called for subordination of one's life to a life greater than ours and therefore for religious worship.

disarray and suffering – was this not an illusion on a par with religious illusions about personal salvation and the afterlife?

Two questions, Russell held, are too often blurred when it
comes to whether human beings can attain happiness: What can
people do to change external conditions such as war, poverty, and
exploitation? And what can a man or woman, here and now, do to
achieve happiness for himself or herself? I see this as a crucial
distinction, too often left out of debates about human chances
for happiness – one that qualifies the ancient advice to set aside
matters that one has no power to influence. In earlier works,
Russell indicated, he had advocated a variety of social reforms in
response to the first question, even gone to prison as a pacifist
in 1918; in his present book he wanted to focus on what people
with sufficient income to secure food and shelter and in reasonable health could do to achieve personal happiness in spite of the
vast social problems still awaiting solutions.[12]

To succeed in such a venture, Russell contended, people ought to
strive to overcome mistaken views of the world, mistaken ethics, and
mistaken habits of life, all of which destroy "that natural zest and
appetite for possible things upon which all happiness, whether of
men or of animals, ultimately depends." Throughout, Russell underscored love and meaningful work as indispensable for achieving the
goal of happiness – the very factors that Freud, too, had long seen as
needed for a healthy personality if not for happiness.

For Freud, however, Russell's distinction between (needlessly)
intractable external conditions and what individuals can do to
achieve happiness in their own lives could not hold. It is civilization itself, he argued, and the necessity to control human
aggression, that inevitably reduce the levels of happiness people
can achieve – a heavy price, he admitted, but one that has to be
paid to constrain the mutual violence that would otherwise break
out. Even as Russell exalted love and work as capable of bringing
lasting individual happiness, Freud now viewed both as inadequate from that point of view: "We are never so defenseless
against suffering as when we love, never so helplessly unhappy
as when we have lost our loved object or its love."[13] And the

pleasures of work are unfortunately accessible only to a few people. Most people work, not to be happy, but under the pressure of necessity. Even for the fortunate few whom Freud believed to be gifted enough to enjoy their work, it "creates no impenetrable armor against the arrows of fortune, and it habitually fails when the source of suffering is a person's own body."[14]

Freud's conviction that misery, on the whole, constituted the human lot could only be reinforced as he witnessed the rise of Nazism and the daily signs that his country, his culture, his civilization, could founder into unspeakable calamity. And his question about the phrase of Plautus – *Homo homini lupus* (Man is a wolf to man) – continues to resonate: "Who, in the face of all his experience of life and of history, will have the courage to dispute this assertion?"[15]

Personal Influences

Freud's and Russell's conflicting views of what people can do to achieve personal happiness go beyond disagreements about the state of the world, as about religion, human psychology, and how best to define happiness. They also reflect the authors' different perspectives on their own lives, influenced by their dispositions and their life experiences and projected, at times, onto all of human existence. In an earlier book, Freud had underscored the difficulty of assessing the influence of purely personal factors when someone pronounces on the future: "the greater or lesser optimism of his outlook on life, as it has been dictated for him by his temperament or his success or failure."[16] To counter any such interpretation of his new book, however, Freud wrote to a friend that he believed he had "not given expression to any of my constitutional temperament or acquired dispositions."[17]

The titles that the two gave to their books highlight the clash between their perspectives. Whereas Russell's title pointed to happiness as capable of being conquered, Freud had intended to call his book *Das Unglück in der Kultur* (Unhappiness in Culture) – then changed *Unglück*, with its meanings of unhappiness, misfortune, bad luck, distress, affliction, woe, calamity, and disaster to the weaker *Unbehagen*, meaning "discontent," "discomfort," "uneasiness." The original title reflected Freud's own sense of *both* bad luck and unhappiness as he neared the end of his life. In earlier decades, he had already gone through prolonged periods of depression, often relieved by opium; now, at seventy-four, he lived with near-constant pain after several operations for cancer of the mouth. Writing to Lou Andreas-Salomé in 1927, he had described his mood as a state of complete disillusionment "comparable to the congealing of the moon, the inner freezing."[18] And in a letter to her written in July 1929 from the mountain resort where he had just finished writing *Civilization and its Discontents*, after disclaiming much importance for a book written so fast, he added: "But what else should I do? One cannot smoke and play cards the entire day; I no longer have the strength for walks, and most of what one might read no longer interests me."[19]

We must keep Freud's desolate condition in mind as we encounter what could otherwise seem to be merely querulous railing at modern ideas and at technological progress – as when he questioned whether the greatly reduced infant mortality and risks of infection for women in childbirth made possible by modern civilization had added to the happiness of humanity. How could he fail to consider the difference for parents who no longer had to grieve for the death of nearly half their children, or for women less fearful of dying in childbirth? Never does his litany seem more personally felt than with respect to the longer life spans that had become possible in his lifetime: As he put it, "And, finally, what good to us is a long life if it is difficult and barren of joys, and if it is so full of misery that we can only welcome death as a deliverer?"[20]

Russell, too, had known periods of severe depression. In an early essay, he had declared happiness unattainable, given the meaninglessness of human fates seen against the indifference of the cosmos. Trapped in what had turned into a loveless first marriage, he had written of the insights from tragedy that make us so awed by the spectacle of death and suffering that we lose "all eagerness of temporary desire."[21] In *The Conquest of Happiness*, however, he diagnosed those who believe (as he once had) that every effort to escape one's miserable lot must end in failure as suffering from "Byronic unhappiness." They attribute their personal misery to the nature of the universe, taking their view to be the only rational one for enlightened people to hold, instead of seeking out the reasons for their unhappiness. His advice to them was to stop brooding about their lot, and to turn, instead, to all the objects of interest and affection that abound in the world: "The secret of happiness is this: let your interests be as wide as possible, and let your reactions to the things and persons that interest you be as far as possible friendly rather than hostile."[22]

Such happiness had seemed within reach for Russell after his second marriage, in 1921, to the scholar and social activist Dora Russell, twenty-two years younger than he – happiness he saw as heightened with the birth of their two children John and Kate. In his *Autobiography*, he looked back at the sheer animal affection he had felt for them, seeing life through their eyes: "The beauty of the Cornish coast is inextricably mixed up in my memories with the ecstasy of watching two healthy happy children learning the joys of sea and rocks and sun and storm."[23]

Dora Russell had published a book entitled *The Right to Be Happy* three years before Bertrand Russell's book appeared. Drawing on her study of French literature and of the challenge from Enlightenment thinkers to theological views relegating happiness to the afterlife, she argued that happiness for all human beings could now be assisted by modern theories of man and the universe. New knowledge about the human body and psychology could make it possible to bring changes in human nature, and social reforms could promote happier societies. Hers was an exultant defense of open marriage and of sexual liberation for men and women alike. She exhorted readers to seize the opportunity to free themselves from the chains imposed by religion, society, and convention and to devise new forms of delight: "Men and women, you have not only a right to such happiness, but the means to this happiness lie ready to your hand. Are they so simple that you must forever pass them by?"[24]

Bertrand Russell fully endorsed his wife's views on marriage and sexuality in his own book *Marriage and Morals*, published in 1928. But by the time he was completing *The Conquest of Happiness*, the extramarital affairs that he and Dora had taken for granted as harmless diversions were devastating their relationship. Their disputes were heightened by the financial problems of keeping afloat the innovative Beacon Hill School they had founded just three years earlier. To make matters worse, their two children were bleakly miserable at that school, even though it was designed to educate them for lives of freedom and happiness.

By 1934, the Russells were divorced. Looking back, Bertrand Russell noted that he had written his book on happiness at a time when he desperately needed to draw on what he had learned from painful experience if he was to "maintain any endurable level of happiness."[25] His own book never went so far as to advocate, as had Dora's, a *right* to happiness – with Thomas Jefferson, he would be content to insist on the right to its *pursuit*; but at no time did he doubt that it was possible to achieve happiness once again. The book shows him sallying forth as jauntily as if he were going to a masquerade, juggling different masks to hold before his face: that of the once unhappy man, now increasingly happy; that of a man fully aware of the burden imposed on human beings by poverty and illness but convinced of the joyful lives some could nevertheless lead; that of the philosopher who has seen through much cant about happiness but also drawn from the wisdom on the subject from Aristotle, Seneca, and others; and that of the social activist and progressive schoolmaster who had understood the real-life obstacles to simplistic hopes even as he cherished such successes as he saw children achieve.

Dogmatic, at times making preposterous claims (as in holding that "the civilization produced by the white races has this singular characteristic that in proportion as men and women absorb it, they become sterile"),[26] Russell's book nevertheless abounds with witty or sensible observations. It conveys the zest that he praised as essential to happiness, as in his admiring descriptions of bold, adventurous men who enjoy shipwrecks, mutinies, earthquakes, conflagrations, and all manner of unpleasant experiences, provided that they do not go so far as to impair bodily or mental health.

In her book *My Father Bertrand Russell*, Kate Tait expresses the sense that many who knew him conveyed, of one who could exhibit just such zest, such adventurousness, at times just the sense of animal happiness that he regarded as too rarely shared by humans. She ends by stating that Russell was "the most fascinating man I have ever known, the only man I ever loved, the greatest man I shall ever meet, the wittiest, the gayest, the most charming"; this in spite of having conveyed in searing terms her unhappiness for most of her life, attributing it in part to his odd educational theories and her own inability to find help from the "recipes" in his books about how to live and conquer happiness.[27]

Freud's youngest daughter, Anna Freud, might well have echoed Kate's glowing characterizations of her father. Freud had chosen her early on as his intellectual partner, one who would be willing and able to carry on his legacy. It is hard to know the extent to which he contributed to her depressed states in childhood and adolescence; but he, like Russell, subjected his daughter to methods meant to liberate her – in Anna's case, taking the step (which would later be ruled out as unacceptable for analysts with respect to their own children) of psychoanalysing her.

Both Freud and Russell were surely strongly influenced by their personal circumstances and temperaments in writing about the prospects for achieving lasting happiness. They could no more avoid such influences than most others who write on happiness. But these influences are far more complicated than some simple contrast of gloomy versus cheery, pessimistic versus optimistic, despairing versus hopeful. Both men were in fact seriously depressed when they wrote; but they drew on their dejection as well as on all the internal resources they could muster to set forth their thoughts on happiness. Russell was also using his reflections in an effort to achieve, once again, the personal conquest of happiness that he still hoped for. Some have characterized his book as deceptive and even dishonest; I think of it instead as representing a huge personal effort that only fully succeeded much later in life with his extremely happy fourth marriage to Edith Finch.[28]

Estimates of Happiness and Unhappiness

What light does recent research throw on Freud's and Russell's conclusions about human chances for lasting happiness? In the first instance, it strongly challenges the two authors' shared conviction that most human beings are unhappy. In global surveys of what people actually say about how happy or satisfied they feel about their lives, the majority of respondents regard their lives as moderately or very happy – at all levels of income and education. The Dutch scholar Ruut Veenhoven, after looking at decades of research on life satisfaction, has concluded that the results flatly contradict Freudian theory: average happiness tends to be above neutral in most countries of the world including some of the poorest areas, such as the slums of Calcutta; when people are asked how satisfied or dissatisfied they are with their lives as a whole, on a scale from 1 (dissatisfied) to 10 (satisfied), averages are most frequently in the 7, 8, and 9 range, whereas Freudian theory would predict averages below 4 everywhere.[29]

Recent research likewise contradicts the assumption, common not only to the two authors but also to many critics of modern civilization such as Rabindranath Tagore and Theodor Adorno, that civilization has somehow added to human misery. People regularly report being happier in societies that enjoy high standards of living, high levels of literacy, good citizenship, individual freedom, cultural plurality, and modernity. Denmark and other Scandinavian countries, along with Switzerland and the Netherlands – societies that all these critics would surely characterize as "civilized" – are regularly found at or near the top of comparative lists.*

---

* Although the surveys document strong preferences on these scores, they clearly cannot resolve religious and philosophical debates about the "discontent" which Freud held to accompany any form of civilization or about the degree to which civilization in its own right promotes or fails to promote human thriving.

In the late 1920s, however, when Freud and Russell wrote, there were no such global surveys, no evidence to go upon save anecdotal observations. Might it be that levels of happiness in many societies including their own were lower in the 1920s than now, and lower still in the nineteenth century, when both authors spent their formative years?

Yes, is the answer given by historian J. M. Roberts. Even after weighing the horrors and mass killings brought about by war and tyranny in the twentieth century, he points to major changes during that century which contributed to higher levels of happiness in many societies. Among these are the unprecedented and revolutionary growth in the world's wealth and the expanding role of democracy and of lifestyles offering much greater choice for individuals. These factors, Roberts says, have made a vast difference in the lives of women, half the human race. He notes that a final outstanding change during the century "seems to be cultural and psychological: the spread of the idea that human happiness is realizable on earth and human destiny manageable."[30] These immense shifts could not have been widely anticipated in 1901, when pain and the threat of infectious disease had to be taken for granted as a given for most people, not least women in childbirth; when infant mortality still ravaged rich and poor families in all societies; and when average life expectancy was close to half of what it is today.

Even at that time, however, there is no evidence that a majority of people were unhappy, still less almost all. Were Russell alive today, he might have revised his estimates of widespread unhappiness. But Freud would most likely have stood his ground, given his conviction that people who think they are happy are in fact deluded and that "No one, needless to say, who shares a delusion ever recognizes it as such."[31] Social scientists who claim to be measuring happiness would therefore, in fact, only be measuring erroneous opinions. Thinkers in the psychoanalytic tradition, such as Slavoj Žižek and Jacques Lacan, still discuss what they see as misunderstandings on the part of individuals and societies when it comes to their own happiness. For Žižek, "Happiness . . . is inherently hypocritical: it is the happiness of dreaming about things we do not really want." As for Lacan's claim that "There is no happiness besides that of the phallus – that is, of the Other," it sidesteps not only happiness research but most other views on the subject.[32]

It is nevertheless useful, the political scientist Grant Duncan suggests, to "amplify our understanding of the politics of happiness" by contrasting contemporary research on happiness to the pessimistic views deriving from the psychoanalytic theories of Freud, Žižek, and Lacan.[33] Psychoanalysts can challenge happiness researchers to probe more deeply the extent to which respondents to surveys do know their own best interests and are aware of the ideological assumptions underlying proposals for increasing collective and individual happiness. The researchers, meanwhile, can prompt psychoanalysts to pay greater heed to what individuals actually say about their own experience and to the corroborating evidence from others who know them well; and, in turn, to take a new look at the evidence on which they base their estimates of general unhappiness, which seemed easy enough to venture in the 1920s but are much harder to maintain today.

While today's findings go against Freud's and Russell's shared convictions that most people are unhappy and that modern civilization contributes to this state, what about the issue on which they disagreed – Russell's claim that "many people who are unhappy could become happy"? Is it possible for people to achieve lasting changes in happiness, unlike the temporary mood swings and ups and downs that all acknowledge? To what extent are we predisposed to experience greater or lesser happiness?

These questions, long debated in the context of views about the sway of fortune or of divine will, even predestination, are now more often phrased in terms of the nature–nurture debate, weighing the roles of heredity and environment. And whereas many secular self-help books and religious writings offer prospects of lasting happiness through personal transformation, researchers are far from unanimous on this score. Two ideas in particular have given pause to many who might otherwise have been prepared to agree with Russell and optimistic self-help books. One, often discussed in terms of a "hedonic treadmill," focuses on the role of adaptation, the other on that of heredity.

The Hedonic Treadmill

The expression "hedonic treadmill" was coined in 1971 by two psychologists, Philip Brickman and Donald Campbell, to link the general concept of adaptation to experiences of pleasure (the Greek *hedone*) or happiness. They suggested that people respond only briefly to the ups and downs in their lives that have to do with income, for example, or health, then go back to a neutral state.[34] Try as we may, we have more in common with squirrels on a treadmill than we might imagine, keeping going just to stay in place.

The phenomenon is well known. In 1762, Oliver Goldsmith quoted an imagined letter from a Chinese philosopher to a slave in Persia, holding that every mind seems capable of entertaining a certain quantity of happiness, which no institutions can increase, no circumstances alter. Gratified ambition, or irreparable calamity, may produce transient sensations of pleasure or distress. "But the soul at length subsides into the level of its usual tranquillity."[35] It is this process of adjusting emotionally to circumstances bringing good or bad luck that psychologists now refer to as "hedonic adaptation."

Adaptation is what allows organisms, from molecules to humans and entire species, to respond to changed circumstances – as when our bodies regulate temperature, hormonal flows, immune responses and much else to achieve a better fit with such changes. We could not survive without being able to adjust in countless such ways, most of them beneath the level of consciousness. At a still higher stage of abstraction, systems, including organisms, attempt to maintain the equilibrium of homeostasis in response to disturbances from the outside. Like all organisms, we humans are programmed to seek to maintain homeostasis to the extent possible.

In 1978, Brickman and two colleagues offered further evidence for the hedonic treadmill. They concluded that lottery winners are not significantly happier, after the initial thrill wears off, than they were before or than those who bought tickets but failed to win; and that victims of accidents resulting in paraplegia turn out to be less unhappy, after a period of time, than most outsiders would have expected.[36] So we respond by some forms of hedonic adaptation not only to adversity but also to *improved* circumstances. Subsequent studies have shown how people accustom themselves to rising income, better housing, or more plentiful food, without finding themselves noticeably happier. Buying a new car or moving to a different climate often brings no lasting increase in happiness. Psychologist Daniel Gilbert and colleagues have compared average human responses to negative news to a "psychological immune system" that helps to ameliorate their experience of "romantic disappointments, career difficulties, political defeats, distressing news." The authors suggest that because people do not take this system into account in looking ahead to the possibility of such events, they regularly imagine that their distress or disappointment will last longer than will actually be the case.[37]

Hedonic adaptation can facilitate the immunity to suffering in a number of ways. Some involve a conscious effort to shift attention away from what is painful or troubling: to look for a silver lining; to make an effort at cheerfulness and optimism; to reflect on what lessons one can draw for the future about how to avoid such experiences or deal better with, say, blindness or deafness; or to redefine, perhaps lower, one's expectations or goals in life. Other forms of adaptation, involving oblivion, denial, compartmentalization, and self-deception more generally, often operate in a near-automatic manner.[38]

Although the study of lottery winners and accident victims was received, in a number of quarters, as proof positive that it would be useless to try to go against the forces of the hedonic treadmill, it also met with critical responses. Economist Richard Easterlin pointed out that the study was based on a small number of individuals and that the paraplegic subjects did not, in fact, return completely to the state of happiness they had enjoyed before their accident.[39] Others have shown that factors such as the loss of employment or the death of a spouse do bring unhappiness for prolonged periods of time. There are also types of suffering impervious to any sort of adaptation for most people: chronic pain, in particular, along with persistent loud noise, and severe depression. As for lottery winners, more recent studies have shown that many actually experience some lasting increase in subjective well-being, in spite of the publicity attending the travails of some winners of exceptionally large amounts.[40] In an article entitled "Beyond the Hedonic Treadmill," psychologist Ed Diener and colleagues examined the hedonic treadmill theory almost thirty years later in the light of continued empirical research. They concluded that it required a number of revisions which, together, "offer hope for psychologists and policy-makers who aim to decrease human misery and increase happiness."[41] In addition, Diener and other psychologists have also focused on the claim that there are "set points" for happiness; and on the suggestion, in the original treadmill theory, that people return to a neutral set point after an emotionally significant effect. To what extent could there be a genetic answer to the questions of why there should be such a high level of hedonic adaptation to good and bad experiences?

Set Points and the Role of Heredity

In 1996, two psychologists, David Lykken and Auke Tellegen, drew on studies comparing several thousand middle-aged twins to argue that about 80 percent of the difference between people when it comes to happiness is determined by heredity.[42] Speaking of "the great genetic lottery that occurs at conception," the authors were "led to conclude that individual differences in human happiness – how one feels at the moment and also how happy one feels on average over time – are primarily a matter of chance."[43]

Such vast and seemingly assured scientifically based claims caught the attention, not just of social scientists but also of the general public, causing considerable consternation. Where human happiness was once seen as affected by the positions of the stars, the interaction of bodily fluids, or the whims of fortune, should we substitute "the great genetic lottery at conception" as playing the largest role in human fates? Just as believers in predestination or astrology have long considered outside influences on human fortunes all-powerful, so many people, confronted with assertions about the role played by heredity, found themselves asking just how much they can realistically expect to do to increase their own chances for happiness. Doesn't the new evidence throw a monkey wrench in the path of anyone setting out, with Bertrand Russell, to "conquer happiness" or of the many self-help books asking readers to "choose happiness"? But doesn't the evidence likewise cast doubt on Freud's theories regarding the powerful influences from early childhood trauma and the Oedipus and other complexes, as well as on behaviorist theories insisting that children can be programmed by early training?

In his 1999 book, *Happiness: The Nature and Nurture of Joy and Contentment*, Lykken announced that he wanted to recant his earlier, pessimistic claim that trying to be happier was as futile as trying to be taller.[44] It had been wrong to be so categorical, he now believed; in fact, people could influence their set points for happiness to some degree. He used the metaphor of "a lake on which our ship is sailing and the higher the lake level, the better we feel" to show how such change was possible.[45] The lake level varies from person to person for genetic reasons, and within persons for physiological reasons. Individuals, he indicated, have quite different "happiness set points"; and people can change their set point somewhat by becoming alert to "happiness makers" – activities and experiences one enjoys – and to "happiness thieves" such as depression, anger, and resentment. Why is that task so much more difficult for those with low set points than for others? Lykken drew both on the original comparisons between identical and fraternal twins, and on findings from studies of both kinds of twins reared separately, to argue that the similarities between the happiness set points of identical twins remained over the decades, regardless of whether they had shared the same home environment or not.

The set point metaphor came into common parlance, with people in many societies wondering what their own happiness set point might be and by what means they could move it upwards.

But the calculations that Lykken and others relied on have not held up in later studies. Estimates of set points turn out to vary a great deal, and experiences such as marriage, divorce, serious illness, and unemployment, contrary to set-point theory, can have long-term effects on happiness.[46] So can political and social changes that bring about lasting increases in felt happiness in a number of societies, not explicable in genetic terms.[47] For instance, levels of well-being gradually rose substantially in the former East Germany after reunification with West Germany, while West German levels remained largely unchanged. As Ronald Inglehart and colleagues argue, in commenting on data from representative national surveys carried out from 1981 to 2007, happiness rose in 45 of the 52 countries for which substantial time-series data were available – changes that simply did not square with the widely held view that it fluctuates around set points.[48]

Neurophysiologist Felicia Huppert likewise challenges the claim that the set point is determined by our genes and that it cannot be changed. She draws on brain research to show how set points shift for individuals and how differently they are experienced, depending on temperament and emotional style. The assumption that our basal level of happiness is primarily under genetic control is inaccurate, she claims, since genes need to be expressed – turned off or turned on. Interventions therefore make it possible to change set points, by modifying external circumstances, attitudes, and behavior. At times, our genes may play the predominant role, but only if we have certain experiences; alternatively our environment may have the greatest influence, but only if we have a particular genotype.[49]

Increasingly, the long-standing tendency to counterpose nature and nurture, heredity and environment, as somehow competing for influence over human fates is coming to be seen as too simplistic. Not only do both obviously play a large role in the development of any living being; what is more important is that they interact continuously. As science writer Matt Ridley puts it in *Nature via Nurture: Genes, Experience, and What Makes Us Human*, genes must be seen as the agents of nurture as well as nature. There is no reason to fear or to disparage the results emanating from genetic research as somehow threatening to human independence:

> Genes are not puppet masters or blueprints. Nor are they just carriers of heredity. They are active during life; they switch each other on and off; they respond to the environment. They may direct the construction of the body and brain in the womb, but then they set about dismantling and rebuilding what they have made almost at once – in response to experience. They are both cause and consequence of our actions. Somehow the adherents of the "nurture" side of the argument have scared themselves silly at the power and inevitability of the genes and missed the greatest lesson of all: the genes are on their side.[50]

Such an approach leaves considerably more scope for human intervention and agency in the pursuit of happiness than those who stress either heredity or environment alone would concede: it rejects altogether the grim prognostications for all of humanity voiced by Schopenhauer and echoed by Freud, to the effect that everything in life demonstrates "that earthly happiness is destined to be frustrated, or recognized as an illusion."[51]

Indeed, Ridley might conclude that Schopenhauer's contemporary, John Stuart Mill, had it just about right in his *Autobiography*. Having fallen into deep despair on realizing that he could no longer share the missionary zeal of his father, James Mill, and of Jeremy Bentham, for promoting human happiness, Mill could not conceive of overcoming his dejection, much less conquering happiness. "I felt as if I was scientifically proved to be the helpless slave of antecedent circumstances; as if my character and that of all others had been formed for us by agencies beyond our control, and was wholly out of our own power."[52] Gradually, however, as he learned to relish poetry and the life of feeling that had been declared out of bounds by Bentham and his father, he came to see that "though our character is formed by circumstances, our own desires can do much to shape those circumstances [so that] we have real control over the formation of our own character."[53]

Three recent books by psychologists respond with a qualified "Yes" to the question of whether people can achieve lasting happiness in spite of the undoubted role played by heredity. In *Authentic Happiness*, Martin Seligman asserts that about half of people's score on happiness tests would be identical to what their biological parents would have scored had they taken such a test, then sets forth a number of choices and activities by which to influence the other half.[54] For example, he has developed exercises, now used in a number of schools, that teach how to express gratitude for positive events in one's life and how to consider forgiveness when it comes to negative events.

The jacket of Sonja Lyubomirsky's *The How of Happiness: A Scientific Approach to Getting the Life You Want*, shows a cherry pie with a large slice taken out – the up to 40 percent of happiness "that is within your power to change."[55] Even though we might view 50 percent of our potential for happiness as determined by heredity and 10 percent by life circumstances, we can do much, she suggests, to influence the remaining 40 percent. We can choose to go against the hedonic treadmill. The title for the first half of the book, "How to Attain Real and Lasting Happiness," would clearly appeal to Russell. But he would surely argue, correctly in my opinion, for doing far more to change the life circumstances (including social conditions) to which Lyubomirsky allocates a meager 10 percent. She emphasizes that it is *intentional changes* in activities and practices that can make a difference in increasing our happiness, indicating differences in how adaptation influences levels of happiness. Adaptation clearly occurs for all changes, but usually does so more rapidly for changes in circumstances such as getting a new car or moving to a different part of the country than for changes brought about by intentional activity, such as starting to exercise or initiating a new project. Among such intentional activities are many which have long been mentioned in self-help books: practicing gratitude and positive thinking, investing in social connections, managing stress, hardship, and trauma, and living in the present.

Ed Diener and Robert Biswas-Diener, in *Happiness: Unlocking the Mysteries of Psychological Wealth*, likewise maintain that individuals, up to a point, can make changes in their lives to add to their levels of happiness. Psychological wealth, the authors argue, is altogether different from having a lot of money. While material resources are definitely helpful, not at all to be rejected as contrary to happiness, psychological wealth is "your true total net worth, and includes your attitudes toward life, social support, spiritual development, material resources, health, and the activities in which you engage."[56] Like Lyobomirsky and Seligman, the authors have a list of what makes for such psychological wealth: life satisfaction and happiness; positive views on spirituality and meaning of life; positive attitudes and emotions; loving social relationships; engaging activities and work; having values and life goals to achieve them; physical and mental health; material sufficiency to meet our needs.[57] Of these, the authors view having loving, or at least close social relationships as by far the most important; and see negative emotions such as fear, hatred, and envy as especially likely to be destructive of happiness.

For such advice, psychologists rely on research findings to show how different activities correlate with average levels of subjective well-being. Readers should keep in mind, however, that *averages* and *correlations* are at issue, in a world of great variety when it comes to what makes people happy. Otherwise those who relish solitude, say, or who thrive even though their outlook on life is anything but rosy, will feel puzzled.[58]

It would also be important for books offering advice on ways to increase one's happiness to provide cautions and qualifications of a moral nature. What if developing greater optimism blinds one to the needs of others? Or if the injunction to be more outgoing and gregarious leads some to become involved in unsavory practices? Advice about how to conduct one's life should be as clear about risks to self and others as are manuals about how to operate an automobile.

Even if caution should be the rule before people simply launch into seeking greater optimism or gregariousness in order to become happier, is not practicing gratitude different? People can be happy without being optimistic or outgoing; but if they can feel no gratitude, they are so lacking in awareness of others as to seem doomed to a thwarted existence. We do need gratitude, and the advice that learning to experience it more fully can add to happiness seems valid for everyone.[59]

As classicist Margaret Visser shows, in *The Gift of Thanks*, we are not born with the ability to experience gratitude or to explain it.[60] It must be learned and practiced, since it requires awareness of others and depends on thinking and attention. Indeed, the words "think" and "thank" have the same root in Old English.[61] To think through what one has to be grateful for is therefore essential to efforts to go against the hedonic treadmill.

Nevertheless, even here moral questions enter in, as Visser shows, drawing on anthropology, myth, history, and literature. Gratitude to whom? For what? And at whose expense – say about inheritance? Just ask siblings who think they have been unjustly treated in a will. These questions can provoke animosities from one generation to the next.

What about gratitude, not to persons but rather to a deity or fate or fortune? For having been born in the first place, perhaps, or being in good health? Or take the example of Socrates, who used to say, according to one source, that he was grateful to fortune for three things: that he was born a human being, not an animal; a man, not a woman; and a Greek, not a barbarian.[62] Similar forms of gratitude are still voiced in many parts of the world. All have to do with feeling lucky to be on one side of some divide. That feeling does add to personal happiness but it bears bitter fruit when it depends upon discrimination against people on the other side.

The personal influences so clear in both Freud's and Russell's outlook on the prospects for lasting happiness may present a special challenge for happiness studies. For while all research requires attention to possible sources of bias, happiness research may have special influence on the mood of some of those immersed in it. Just as prolonged study of child abuse, say, or torture, can have a depressing effect on the researchers themselves, at times to the point of "burn-out," so engaging in the study of happiness can induce greater optimism about ways to enhance it. What is more natural, then, than to speak in general terms without pausing to discuss individual differences from averages or the moral questions raised by pursuits of greater happiness? The risk, in so doing, is to offer advice so sweeping and unquestioning as to cry out for challenge from a Montaigne, an Elisabeth of Bohemia, or a William James.

* * * *

Reading Russell's and Freud's books side by side also offers perspective on the vexed questions surrounding claims and counter-claims about illusion: how best to understand the deluded sense that people can have regarding their own happiness; how best to respond to what they say about their felicity; how to view the conflicting political and religious claims about true versus illusory happiness; and whether forms of illusion might be not only inevitable but also beneficial. Both thinkers challenge the assumption that whatever people say about their happiness has to be taken at face value. But because both use the language of illusion and delusion so freely to dismiss the views of those who disagree with them, they leave themselves open to the same criticism. Why shouldn't a Christian, for example, reject Freud's or Russell's atheism as similarly deluded? Among their contemporaries Simone Weil asserted as indisputable the fact that "there is a reality beyond this world, outside space and time, outside man's mental universe."[63] In all the crucial choices in human existence, she wrote, the only choice is between supernatural good on the one hand and evil on the other. She could have been throwing down the gauntlet to Freud and Russell in claiming that

> If a captive mind is unaware of being in prison, it is living in error. If it has recognized the fact, even for a tenth of a second, and then quickly forgotten it in order to avoid suffering, it is living in false-hood. Men of the most brilliant intelligence can be born, live, and die in error and falsehood.[64]

VIII

ILLUSION

True and Illusory Happiness

Happiness is to rejoice in you and for you and because of you.
This is true happiness and there is no other.

St Augustine, *Confessions*[1]

Religion is the sigh of the oppressed creature, the sentiment of a
heartless world, and the soul of soulless circumstances. It is the
opium of the people. The abolition of religion as the illusory
happiness of men, is a demand for their real happiness.

Karl Marx, "Toward a Critique of
Hegel's Philosophy of Right"[2]

IT IS NATURAL, GIVEN THE VAST CONSEQUENCES THAT ADOPTING
one view of happiness rather than another can have for individual
lives and for institutions, that some should wish to single out what
constitutes "true" or "real" happiness. For many in politics or religion it comes to matter utterly to believe that one view of happiness is the only correct one and to warn against the snares and
delusions of those who peddle different perceptions.

Few proposals for how to attain true or real happiness fail to
target competing beliefs as illusory, blind to true reality, often
intended to lead the gullible astray. The more convinced people
are that they know what true happiness is, the less likely they are
to find value in examining conflicting views. Instead, they might
respond to challenges by saying: "Well, of course you can call
anything happiness, but this only goes to show how far you are

from understanding the true meaning of the term." Those who aim to dispel the false consciousness or other illusory beliefs that they see all around them have held skeptics to be at best self-deluded, at worst in league with agents of deceit.

For some who prescribe a specific path to true happiness, diverging beliefs are not only false, not only deluded, but also dangerous. Teresa of Avila, seeing them as temptations from Satan, offered step-by-step directions for reaching true or genuine or perfect happiness, following the path she herself had taken. Others set forth different paths. This is how the Islamic mystic al-Ghazali put it:

> For perfect happiness mere knowledge is not enough, unaccompanied by love, and the love of God cannot take possession of a man's heart till it be purified from the love of the world, which purification can only be effected by abstinence and austerity. While he is in this world a man's condition with respect to the Vision of God is like that of a lover who should see his beloved's face in the twilight, while his clothes are infested with hornets and scorpions, which continually torment him. But should the sun arise and reveal his beloved's face in all its beauty, and the noxious vermin leave off molesting him, then the lover's joy will be like that of God's servant, who, released from the twilight and the tormenting trials of this world, beholds him without a veil.[3]

What if people ask for evidence of such a state of perfect happiness? Many who, like Teresa or al-Ghazali, propound a prolonged series of steps of initiation or preparation, answer that *true* happiness cannot be perceived and fully experienced until the end of such a process. For them, even posing questions about evidence shows how mired you still are in illusion. However strongly the advocates of perfect or real happiness differ with one another, they are as one in holding that once you achieve enlightenment or insight, you won't have to ask any more questions.

That point may come only in some future state of society, as for Marx – in "a higher phase of communist society, when the enslaving subordination of the individual to the division of labour, and with it the division between physical and mental labour, has vanished."[4] Or else it may come in the afterlife, as for Aquinas, who declared that only then will perfect happiness be achieved: "Final and perfect happiness can consist in nothing else than the vision of the Divine Essence."[5]

Buddhist texts rarely depict perfect happiness so graphically. Buddha did not himself link the attainment of nirvana, or bliss, to visions of a personal God or of the afterlife. Nor did he even define nirvana; he held, rather, that the noble eightfold path of morality would lead to its attainment through elimination of ignorance and selfishness.[6] But Buddhists, too, characterize ordinary conceptions of happiness as illusory: as tempting humans to ignore the state of samsara in which they live, beset with ignorance and suffering, caught up in an endless cycle of birth, death, and rebirth unless they strive for escape from this cycle through attaining enlightenment.[7]

Advocates of some single path to true happiness are not the only ones to dismiss other views of happiness as illusory. Skeptics, far from accepting the accusation that they live in illusion, may cast it back upon believers in such a path. From earliest times, skeptics who have questioned what they saw as unverifiable dictates regarding happiness have sometimes specified as illusory precisely those dictates. In ancient India, for example, adherents of the materialist doctrine of Cārvāka warned that the Hindu beliefs regarding gods and religion and immortality were illusions propounded by priests to deceive the people. All would fare better if they recognized that pleasure and pain were the central facts of life.[8] In China, around the same time, Confucius took an agnostic stance regarding what might await us after death, refusing to speculate: "You do not yet know life, how could you know death?"[9] A few centuries later, in Greece, Epicurus spoke out against the superstitions that caused what he saw as unnecessary anguish among his fellow Greeks about what would happen after death. Fears generated by accounts of Hades in myths and heroic tales were needless, he argued. Once dead we experience nothing; to think otherwise is to live in illusion. We should clear our minds of such imaginings and focus on what brings happiness in the here-and-now.

Doctrines about large-scale illusion and delusion draw their force from our everyday experience of mirages and other illusions. We know from experience what it is like to be wrong, even deluded, about what we later see is a reality that stared us in the face. We recognize the stock figure of the person living in a fool's paradise, in the spell of what seems unshakeable illusion. And we can share the dilemma for outsiders of whether, and if so how, to intervene so as to bring about an awakening to reality – rarely more painful than when the deluded person rejects all meddling with circumstances felt to be supremely blissful.

Debates about what to do under such circumstances differ in fundamental ways from the political and religious arguments about what constitutes true or real happiness; the issue here relates, rather, to questions about the boundary between legitimate and illegitimate interference by outsiders in what they see as people's deluded ways of living. "What if we cherish what you dismiss as our illusions?" they might respond. "What if we think such experiences more 'real' than anything you have to prescribe? What if we have no desire to mature along the lines you suggest, and if your efforts to help us by dispelling our illusions turn out, instead, to harm, perhaps to destroy our happiness?" The story of how the Greek merchant Lycas was "cured" of his illusory bliss contrasts the perspective of someone relishing his illusion with that of outsiders wondering how to break through to him. It was already ancient history in 19 BC, when the Roman poet Horace referred to it in his Epistle to Julius Flores.

The Deluded Bliss of Lycas

Horace tells of how the wealthy Greek merchant Lycas of Argos would spend long hours in an empty arena on his estate, delighting in seeing imagined performances of his favorite tragedies, much to the dismay of all around him. How in the world could he persist in a belief, however blissful, that everyone knew to be utterly false? Yes, he was kind to his wife and servants; but he was in effect forcing them to choose between going along with what they saw as his madness or intervening to restore his sanity. When his relatives finally managed to drive out the disease by powerful doses of hellebore, Lycas cried out in despair at his loss: "*You* call it rescue, my friends, but what you have done is to murder me! You have destroyed my delight and forcibly swept away from my mind the most gloriously sweet of illusions!"[10]

Who was right – Lycas or his relatives? The story came to serve as a familiar thought-experiment: an imagined test case stripped down so as to isolate specific circumstances that sharpen the contrast between reality and illusion, much as Robert Nozick's example of the Experience Machine does today. But whereas the story of Lycas allows listeners to empathize not only with Lycas but also with his bewildered family and friends, Nozick asks individuals to decide for themselves about choosing to enter a state of continued happiness completely disconnected from ordinary existence. The choice is not whether or not to wish to *remain*, like Lycas, in a state of blissful illusion, but whether to enter into such an existence at all, while fully aware of its illusory nature.

In Lycas we have a man experiencing great happiness and thoroughly satisfied with his life, but deluded in the eyes of everyone around him – a state clearly different from the transient fantasies and misconceptions of everyday life. By conveying his shock and misery when he was "recalled to reality," the story highlights the difficult moral conflict such cases can present to people wondering whether to intervene, to turn a blind eye and hope for the best, or to serve as what are now called "enablers" of, say, alcoholism or incest. In turn, the story raises moral questions about the freedom to seek happiness without interference from others. On this score, commentators often sided with Lycas.

Erasmus, in *The Praise of Folly*, refers to "that Greek in Horace . . . whose mania took the form of going to the theater every day imself, to laugh, applaud and cry 'Bravo' because he thought wonderful plays were being performed, though in fact nobody was there at all."[11] Why treat such a merry and harmless folly as some sort of disease, to be expelled by medication? The pleasure of this delusion, Erasmus concludes, is but a foretaste of the unspeakable bliss in Paradise promised to those who love God. Persons who experience such madness while still on earth may laugh and cry intermittently and speak incoherently. When they awaken, "they say they do not know where they have been, whether in the body or out of it, awake or asleep. They cannot tell what they heard, saw, said, or did, except through a mist, as in a dream. All they know is that they were supremely happy while they were out of their minds, and they regret their return to reason because their one desire is to be continually mad with this sort of insanity. And yet this is just the tiniest taste of bliss to come."

Montaigne takes a different view in considering one who "thought he was perpetually at the amphitheatres watching entertainments, spectacles, and the finest comedies in the world." Many philosophers might be of Lycas's mind, Montaigne held, cherishing illusions that bring tranquility to one's existence where "strong live reason is not powerful enough."[12] But even as he rejected as excessive the claim attributed to Socrates by Plato, that the unexamined life is not worth living, Montaigne did insist that we ought nevertheless to try to know more about ourselves, engulfed in mists of illusion and error as we may be.* By itself, knowledge surely cannot bring happiness; yet neither can self-deception. The bliss of delusions such as that of Lycas is not, after all, worth having.

---

* Indeed, there is no reason to believe that either Socrates or Plato regarded the lives of fellow citizens who neglected such self-scrutiny as somehow not worth living. Socrates, charged with impiety and corrupting the young, was defending his methods of inquiring into how best to live. He stated that the best he could do was to let no day pass without discussing goodness and examining both himself and others; then immediately pointed out to his fellow Athenians making up the jury that he knew they would not believe this – taunting them, as it were, challenging them to condemn him to death. (Plato, *Apology*, 38a.)

Alexander Pope brought doctors into the story – "the damn'd Doctors and his Friends" who immured Lycas: "They bled, they cupp'd, they purg'd; in short, they cur'd."[13] Pope's pointing to the outsiders' zeal to intervene and the devastation that doctors can wreak in trying to "cure" seemingly happy individuals resonates today: do individuals have the right to resist psychiatric treatment, and when, and with what safeguards, is it necessary for outsiders to intervene?

A related controversy has to do with whether or not illusion more generally contributes to happiness. "Know Thyself," one of the two mottoes inscribed on the temple of Apollo at Delphi, was contested from the outset. There was, after all, a second motto inscribed on that temple: "Nothing to excess." The first motto signals the impossibility of understanding and responding to the rest of one's world without first understanding oneself, even as the second warns against distortions such as those arising from excessive self-scrutiny.

Thinkers as different as Confucius, Buddha, Montaigne, and Kant have stressed the need to seek self-knowledge in order to mature. The eighteenth century was traversed, as Robert Mauzi puts it, by a grand Stoic dream, shared by Montesquieu, Voltaire, Diderot, and Rousseau, of the slow ascension from illusion to truth and in turn to true happiness.[14] But illusion had long had its defenders as well. Erasmus was one of many who cautioned that self-knowledge could be unbearable without soothing illusions about ourselves. And Francis Bacon asked whether anyone doubted "that if there were taken out of men's minds vain opinions, flattering hopes, false valuations, imaginations as one would, and the like, but it would leave the minds of a number of men poor shrunken things, full of melancholy and indisposition, and unpleasing to themselves?"[15] Julien Offray de La Mettrie, himself a physician, believed that it was to do a disservice to cure anyone in delirium and madness – why not leave them in whatever agreeable dream they cherished?[16] But none spoke up for illusion as eloquently as did the Marquise Émilie du Châtelet, a mathematician and physicist who dared to do so from within the very camp of the Enlightenment *philosophes*.

Madame du Châtelet and the Defense of Illusion

To be happy, it is necessary to have done away with prejudices, to be virtuous and in good health, to have tastes and passions, and to be capable of illusions; for we owe most of our pleasures to illusion, and unhappy are those who lose it. Far, then, from seeking to dispel illusion with the lantern of reason, let us try to thicken the glaze that it places on most objects, more necessary for them than are the cares and adornments for our bodies.

Madame du Châtelet, *Reflections on Happiness*[17]

To better understand the passionate debate that the subject of illusion could inspire, consider the relationship between Voltaire and Madame du Châtelet in the 1740s. At thirty, she had left her husband and her children behind in Paris to go to live with Voltaire, beginning with a stay at her nearby château of Cirey. There, the two experienced what both held to be unequaled happiness, relishing one another's company in its serene libraries and beautiful parks; and directing productions in the estate's small theater where two troupes put on comedies and tragedies including their own. Looking back at that happiness seventeen years later, Madame du Châtelet wrote of her efforts to prolong it despite Voltaire's infidelities and increasing coldness, in her remarkable *Reflections on Happiness*. "My heart, unsuspecting, enjoyed the pleasure of loving and the illusion of believing oneself loved."[18] At the very time when her friends Diderot and Voltaire and all who were preparing the *Encyclopédie* were combating what they saw as the illusions of unreason and superstition, and upholding the rule of reason, she took up the defense of a modicum of life-preserving illusion in order to achieve greater personal happiness. For Robert Mauzi, the originality and personal tone of her essay stand in stark contrast to the many dozens of dreary eighteenth-century treatises on happiness: hers alone is moving, he argues, because of its self-revelation and its hints at suffering first mastered, then transformed.[19]

Madame du Châtelet knew how sharply her claim that a happy life calls for passions and illusions clashed with the priority that

Voltaire and her Enlightenment friends gave to reason. She was as leery as they of what she called the "prejudices of religion" such as those to which they felt Pascal had succumbed. To be sure, she agreed, the lantern of reason should be held up to beliefs such as that earthly happiness is illusory compared to hoped-for felicity in the afterlife. But she took issue with those arid souls for whom a life of reason was all that was needed. Reason should never be marshaled to dispel the illusions that bring personal happiness.

Foremost among the illusions she prized was the belief that the years of passionate love between her and Voltaire had not ended. When that belief finally became untenable, she found that two other passions, for study and for gambling, offered great solace. The first, her lifelong enthusiasm for classical literature, modern languages, and the sciences, she explained, was, of all passions, the one that made her happiness least dependent on others. According to Voltaire, she was working to finish translating Isaac Newton's *Principia Mathematica* into French up to the very moment of giving birth to a baby in 1749; she died, forty-three years old, shortly thereafter; but her translation survived. It remains the standard French rendition of Newton's work.[20]

It was the second passion, however, the bliss she derived from gambling, that brought her the greatest excitement, born of the hope and fear that love had once generated: hope that a stroke of luck was around the corner and fear of the great financial losses she might – and did – incur. Philosophers wrongly call the passion for gambling irrational, she said; it need not be thus for anyone who keeps it under control and knows how to postpone it to their old age – exactly what she herself could never do. Voltaire, who saw her gambling away fortunes, regarded her as utterly self-deceived about the harmlessness of her passion. At one point he had to declare publicly that he would not be responsible for paying her debts.[21]

In her essay, Madame du Châtelet insisted that such illusions don't deceive us, any more than do optical ones or the fictions we enter into so wholeheartedly at the opera. And just as we can refrain from going backstage to see the wheels and machines that make it possible for figures to seem to be flying on stage, so we need not confront the illusions that are among the great machines driving happiness in our own lives. But such a comparison between optical illusions and illusions about gambling, Voltaire might well have answered, obscures the risks associated with the latter. Optical illusions, much like the "suspension of disbelief" by readers of novels or audiences at the opera, involve our conscious shifting between different perceptions. And *trompe l'oeil* paintings expressly intend to trick viewers into first believing that images are real, then taking delight in seeing that they are not.[22] Such is hardly the case with illusions about one's chances of winning at the gambling table.

Psychologists on Positive Illusions

Just as people's beliefs about their health matter far more to their sense of well-being than their actual health, so the same disconnect often exists when it comes to their beliefs about their own intelligence, attractiveness, and moral excellence. Americans now have a term, the "Lake Wobegon effect," based on Garrison Keillor's radio series *A Prairie Home Companion*, about a town where "all the women are strong, all the men are good-looking, and all the children are above average." Psychologist Shelley Taylor, examining what she calls "positive illusions," holds that people have at least three mildly self-enhancing beliefs: "self-aggrandizing self-perceptions, an illusion of control, and unrealistic optimism about the future."[23] According to Taylor, individuals who manage to maintain such positive illusions about themselves are more rational, successful, creative, as well as happier and healthier – even more likely to live longer than those less given to such fictions.[24] In a 2005 article, she describes an early study of men who had been diagnosed with AIDS and were expected to die. Those who held on to their optimistic beliefs despite the progression of the disease lived an average of nine months longer than those who had accepted their decline and were "preparing for their inevitable death."[25]

Others have challenged some of Taylor's claims.[26] But few doubt that optimistic illusions can be beneficial at times of crisis. Sometimes it is vital to be able to shield oneself from information one cannot deal with, as in taking extra time to digest frightening news about illness or a family member's death – what physicians call "pulling down the curtain." It is when self-shielding no longer merely filters but distorts or blocks out needed information about dangers we could otherwise avert that it can do great harm. Far from protecting individuals, such defenses then leave them unaware of threats they could have tried to avoid or overcome, as when people block out, until it is too late, knowledge of the symptoms of a disease that is curable in its early stages.[27]

Is it possible to distinguish beneficial illusions from harmful ones, such as the belief that one runs no risk from high-speed driving or that one is invulnerable to dangers from substance abuse? Taylor has distinguished between positive illusions and defense mechanisms such as repression and denial: the latter alter reality whereas illusions simply reinterpret it in the best possible light.[28]

Such distinctions, however, are precisely those that are hardest to make for individuals living with illusions, whether positive ones or not. Madame du Châtelet, for instance, caught in what would now be diagnosed as addiction to gambling, regarded hers as the most benign and cherished of illusions. Some people clearly exercise better judgment than others. But as psychiatrist Harry Stack Sullivan pointed out, speaking from personal experience with alcoholism, the powers of denial are especially strong for individuals at greatest risk. He described the varieties of "selective inattention" to warning signals as "the classic means by which we do not profit from experience which falls within the areas of our particular handicap."[29]

Recent studies of smokers show how they use such means to block out awareness of risks. They may believe that such risks are for others, not themselves, or that risks will be reversed after they stop smoking later in life.[30] There is no evidence, here, of any distinction between positive illusions and other defense mechanisms.

Psychologist Martin Seligman, though he would scarcely encourage false beliefs about smoking or gambling, suggests that illusion can enhance happiness about love. He advises lovers to keep their illusions, citing studies showing that couples in which both parties maintain illusions about one another are happier than others. Remarkably, "the bigger the illusion, the happier and more stable the relationship."[31] And Seligman points to a study of optimism and pessimism in marriage: "The upshot of this is straightforward: Optimism helps marriage. When your partner does something that displeases you, try hard to find a credible temporary and local explanation for it: 'He was tired,' 'He was in a bad mood,' or 'he had a hangover,' as opposed to 'He's always inattentive,' 'He's a grouch,' or 'He's an alcoholic.' "[32]

To such a rosy picture of what helps marriages, Sullivan might suggest prudence. He might challenge Seligman's claim that "the bigger the illusion, the happier and more stable the relationship"; and ask: How easy is it really, given the vigilance of selective inattention, for people to decide when explanations are "credible" or when healthy skepticism should moderate optimistic illusions? What if your partner really is an alcoholic? Or engages in domestic violence? Or, like Madame du Châtelet, gambles away much of what she owns, oblivious to the needs of her children, borrowing from Voltaire and many others and still unable to stop?[33]

Personal accounts by family members of gamblers, of swindlers, and of child abusers corroborate what studies show about how the many shadings of illusions, selective inattention, denial, and compartmentalization affect both the wrongdoers and the family members who serve as enablers. What are the moral responsibilities of those caught up in such circumstances? And, more generally, when do what appear to be harmless, possibly beneficial, illusions risk endangering others? The literature concerning positive illusions focuses strictly on the possible benefits for the individuals who harbor such illusions: for their mental health, their psychological well-being, their ability to cope with illness; but not on possible effects on others – on family members, perhaps, or colleagues or health care professionals.

Ignorance, Self-Deception, and Moral Choice

> Every man, wherever he goes, is encompassed by a cloud of comforting convictions which move with him like flies on a summer day.
>
> Bertrand Russell, *Sceptical Essays*[34]

Human beings live surrounded by innumerable sources of erroneous belief. Illusion, self-deception, errors, and biases of all kinds interact in ways that may be impossible to disentangle. Mirages may deceive us; our eyes and ears deceive us all the time. Following Georg Simmel, William James, and Sigmund Freud who pressed the exploration of illusion – its ubiquity, its uses, its allure, and the risks it can present – psychologists have studied the role of error and misunderstanding with respect to subjective well-being. In his book *Stumbling on Happiness* Daniel Gilbert reports on experimental studies that underscore the surprising degree to which people not only misperceive the present, but misremember the past and mis-imagine the future.[35]

Self-deception includes all the ways in which individuals are thought of as contributing to this state in themselves. It need involve no conscious intent to deceive and interacts with ignorance and misjudgment and the many ways in which we can misunderstand what is communicated. Jonathan Swift stresses the link between self-deception and happiness, with the Lycas story in mind. If we ask, he writes, what is generally understood by happiness, "we shall find all its properties and Adjuncts will herd under this short Definition: That, it is a perpetual Possession of being well *Deceived*."[36] Such a state constitutes the "sublime and refined Point of Felicity . . . The Serene Peaceful State of being a Fool among Knaves."[37] Swift included, in the category of "Knaves," those conquerors and propagators of religious and philosophical creeds whose imagination has kicked their common sense out the door: "the first Proselyte . . . [such a man] makes is Himself, and when that is once compass'd the Difficulty is not so great in bringing over others; a strong Delusion always operating from *without*, as vigorously as from *within*."

We learn from neuroscientists how some shutting out of information occurs as reflex responses over which people have no control whatsoever. Had we no such protection from all the sensory impressions that bombard us every second, we would be exposed to far more than we could cope with: it would be like having no skin. We deal with the vast majority of these impressions at the simplest organic levels; against others we mobilize myriad forms of rationalization, denial, and other psychological defenses.

Brain scientist Eric Kandel explains the role of the signals that operate at the neurological level from the most primitive organisms all the way to humans. The neural circuits use specific molecules, conserved through millions of years of evolution, to generate signals within and between nerve cells. A signal from a presynaptic neuron produces a "change in the membrane potential of a post-synaptic neuron."[38] Kandel quotes William James on how his attention singles out certain signals from amidst millions of items that are present to his senses without properly entering into his conscious experience: attention "is the taking possession by the mind, in clear and vivid form, of one out of what seem several simultaneously possible objects or trains of thought. . . . It implies withdrawal from some things in order to deal effectively with others."[39]

Such attention, Kandel holds, gives certain signals *salience* in the midst of countless others. Sometimes the shift occurs unexpectedly: "I will put my foot on the brakes automatically when a car suddenly cuts in front of me while I am driving on the road." Understanding the role of salience is crucial in sorting out how denial and other forms of self-deception can deflect or distort signals; and how drivers, for instance, can train themselves to respond more quickly to particular types of danger signal.

Once we recognize the extent to which signaling occurs within organisms as well as between them, we have a point of departure for examining the role of intention, belief, and moral choice when it comes to people's ending up deceived and sometimes deluded about happiness. These roles are especially difficult to untangle when it comes to brain signals within organisms of which individuals are only partially aware. Take the signal sent within the brain of someone addicted to alcohol or tobacco, as the demand builds up for pleasure and for at least temporary relief from craving. To say that one part of an addict's brain is intentionally deceiving another makes little sense. At the same time, the question about what addicts believe about themselves and the choices they confront requires asking about the extent to which they participate in deceiving themselves.

Philosopher Barbara Herman has written of "rules of moral salience" in *The Practice of Moral Judgment;* and of taking seriously the question of how we might become more clear about personal responsibility, learn to become more aware of moral choices *before* problems overwhelm us.[40] Such moral salience has elements in common with neurological salience as described by Eric Kandel. Just as drivers can learn to respond to the salience of danger signals in traffic, so we can learn to become more alert to dangers of a moral nature – say of being drawn unwittingly into acts we would have wished to steer clear of had we recognized in time how they would destroy our happiness. People who find themselves accused of complicity in financial fraud, for example, often wonder, belatedly, why they had not seen the problem in time, but rather clung to comforting illusions about how all was going to be for the best.

Moral salience calls for the development of the balance of empathy and resilience that I have stressed in earlier chapters, so as to be alert, in confronting new situations, to signals that might otherwise go unheeded. The moral awareness with which people choose to pursue happiness matters from the point of view of the kind of person they want to be, and of risks to self and others that they would want to recognize before it is too late.

To avoid adding to such risks, I suggest, greater alertness to the role of ignorance and self-deception in moral choice is indispensable. This is not to reject the pleasures of daydreaming or imagination, nor to ignore the fact that resilience may require shielding ourselves from information we cannot digest right away. On the contrary: the energy required to overcome the adaptive impulses that dull our awareness generates new opportunities for imaginative, timely responses to moral risks. But we struggle along with such thick layers of bias and rationalization and denial that our choi-

suffer immeasurably; we have every reason to resist complacency about them.

I take it that Iris Murdoch had similar alertness in mind in asserting that the chief enemy of morality is "the tissue of self-aggrandizing and consoling wishes and dreams which prevents one from seeing what there is outside one."[41] These attack morality at its roots by skewing how we perceive ourselves and others, what we think and feel, and what we choose to do and to be. To mature, she suggested, we have to learn to navigate "through a continuum within which we are aware of truth and falsehood, illusion and reality, good and evil. We are continuously striving and learning, discovering and discarding images."[42]

One way to learn to navigate in this way is by thinking through moral problems encountered in fiction, autobiographies, and thought-experiments such as that of Lycas and his family; or in case histories, from those debated in Stoic, Confucian, and religious traditions of casuistry to those discussed today in classrooms or in the media. Each time, we can imagine encountering such problems ourselves, but have the freedom to sort through them as an observer rather than as a participant. For one of countless examples that can serve such a purpose, consider business consultant Barbara Ley Toffler's account of meeting with a group of young managers at Arthur Andersen in the wake of its involvement in the Enron scandal.[43] She asked them to discuss how they would have responded if one of the partners had requested them to do something they knew was wrong. All squirmed. No one contradicted the one who finally answered that he would have gone along. Why? "Partners don't want to hear bad news." Had the young managers stopped to think of their own responsibility as well as their long-term satisfaction with their lives, they might have been less willing simply to cater to the partners' unwillingness to confront reality by hearing bad news.

The same is true in every walk of life. Colleagues and students often refrain from intervening when physicians, impaired by alcohol, depression, or substance abuse, endanger patients.[44] The danger is compounded whenever therapists who have optimistic illusions about their own competence undertake to "help" patients maintain *their* illusions. Studies have found that the least competent individuals are especially likely to cherish such benign beliefs about themselves. According to one review of studies of why the most incompetent people are the ones least likely to know this fact about themselves, "poor performers grossly overestimate their performances because their incompetence deprives them of the skills needed to recognize their deficits."[45] Because such individuals don't recognize how poorly they have done on tests or in real-life situations, they are also least able to benefit from critiques or other kinds of feedback.

In reading contemporary philosophers on happiness, I have the distinct impression that most of them are less sanguine than many psychologists when it comes to the benefits of illusory happiness. A frequent thought-experiment in philosophers' discussions involves someone believing that his marriage is blissful, his spouse loving, and that his colleagues at work respect him; only none of it is true. In what way would it be better for him to know the miserable facts about his life? Perhaps, like Lycas, the individual at the center of the web of illusion would reject all interference, mobilizing every form of self-deception in the process. While philosophers disagree about whether we could rightly call this man happy, most agree that they themselves would want to know the facts. James Griffin says, in *Well-Being: Its Meaning, Measurement, and Moral Importance*, that he prefers, "in important areas of my life, bitter truth to comfortable delusion . . . because it would be a better life for me to live."[46] Wayne Sumner, stating that most of us will surely agree with Griffin in preferring reality to illusion in some key sectors of our lives, points out that we often collude in protecting our carefully safeguarded belief structure, knowing it to be fragile: "deception then shades into self-deception."[47]

The shading of deception into self-deception speaks to the obstacles that even the happiest of illusions present for perceiving one's moral responsibilities, and to the importance of striving for greater alertness to them. There are no watertight compartments between the self-deceived states in which individuals live and how they respond to others, in their families, say, or at work, to take the example of the man so ignorant about the affection of his wife and the respect of his colleagues. How one views one's life is bound to affect how one treats others. To be blind to the most basic realities in one's life is to be rudderless when it comes to dealing with even the closest intimates.

This is one reason why I find Immanuel Kant's linkage, in the last book he published, of dishonesty toward others and incomplete or twisted honesty with oneself as persuasive as ever. If one wishes to lead a principled life, he argues, "the highest maxim, uninhibited truthfulness toward oneself as well as in the behavior toward everyone else is the only proof of a person's consciousness of having character."[48]

Kant's stress is on *truthfulness* rather than on some unachievable full *truth*. The fact that we can strive for the former, no matter how far out of reach the latter may be highlights the human potential for choice in communication with others as well as inwardly. In view of what we now know about the extent of memory distortion and the brain mechanisms whereby we filter information, I believe Kant might have amended his view so as to speak, not simply of uninhibited truthfulness with ourselves and others but of *striving* for such truthfulness and above all of not inhibiting it still further. But he surely would have continued to insist that such efforts represent the only way we have of knowing whether we care about ourselves as responsible moral agents. And he, who had defined happiness as uninterrupted and lifelong consciousness of pleasure, and thus impossible to achieve on earth, might well have been prepared to agree with Mencius, writing two thousand years before him, about being true to himself as his highest *joy*, then linking that self-knowledge to the Golden Rule:

> All the ten thousand things are there in me. There is no greater joy than to find, on self-examination, that I am true to myself. Try your best to treat others as you would wish to be treated yourself, and you will find that this is the greatest way to benevolence.[49]

IX

THE SCOPE OF HAPPINESS

ANYONE SETTING OUT TO EXPLORE WHAT WE CAN LEARN ABOUT the role of happiness in human lives encounters a daunting multitude of reflections, analyses, and flights of the imagination, of experiences of happiness and of happiness only longed for. They convey the subject's remarkable scope, in two senses of that term: extent, range, reach, sweep of perceptions, thoughts, or actions; and freedom, latitude, and leeway.[1] I see both meanings as relating to each of my two aims in this book: that of bringing together writing and thinking about happiness by philosophers, poets, religious thinkers, and others with research by natural and social scientists; and that of examining, against this background, the limits imposed by perennial moral issues about how we should lead our lives and how we should treat one another.*

Many students of happiness discussed in these pages have illuminated its scope in the first sense of the term. Some have felt free to range across cultures and centuries and disciplines to draw on literature from earliest times, on history, philosophy, religion, the arts, and, increasingly, the sciences. Others have examined particular problems related to happiness in depth. Together, they allow us, in turn, to reach beyond our own perspectives to appreciate

---

* "Scope" derives from the Greek *skopos*, meaning "one who watches," but also the object of attention such as a "goal," "aim," or "target"; and in turn from the Indo-European root *spek-* that can be found in words such as "spectrum," "spectacle," "aspect," "introspect," "perspective," "respect," "retrospect," and, via the Greek *skeptesthai*, "to examine," found in terms such as "skeptic."

the full scope and the extraordinary variations in how people perceive and experience happiness.

Without looking to personal accounts of the experience of happiness or to works of literature, it would be impossible to take its full measure, to envisage its complexity and its variability and nuances. Philosophical analyses and reflection since antiquity are likewise indispensable in extending and deepening our understanding; and thought-experiments, in turn, challenge intuitive assumptions about happiness. The social sciences and neurosciences test such assumptions as well, bringing new empirical methods to study the factors that contribute to, or detract from, happiness.

I have stressed, throughout, the importance of drawing on *all* the resources now available for exploring happiness, and of learning to shift perspectives in so doing. I have pointed to the risks of making do with unreflective, stereotyped conclusions about happiness, referring to what the literary critic I. A. Richards called "premature ultimates" that bring investigation to an end too suddenly.[2] The many biases and failures of perception that fall under the heading of premature closure include tunnel vision, cramped thinking, and the short-circuiting of debate.

In medicine, the term "premature closure" is used to denote what happens when fibrous joints between the bones of a baby's skull close too early, risking deformation and stunted brain development.[3] The medical use of the term offers a telling metaphor for such closure more generally, whether it be for short-changing or bypassing altogether efforts at the fullest perception and the fairest, most careful deliberation, or settling for uncritical acceptance of simplistic choices and inadequate implementation. In debates about happiness, I believe that serious attention to the scope and depth of the many promising forms of research now under way, in the light of past reflection, can help counter the ever-present temptations to arrive at premature closure.

Take the once-common notions that genius and melancholy go together. We now know a great deal more about different kinds of melancholy and types of depression and how each relates to creativity, including genius. The same is true of the more recent supposition that persons who cherish solitude are most likely neurotic. The fact that a proportion of people who lead solitary lives also exhibit traits that researchers associate with neuroticism should never be allowed to obscure the fact that others do not. Correlations based on averages often conceal wide variations among individuals.

Another trend that I see as evidence of premature closure is that of vaunting resilience as a strictly valuable trait, without distinguishing the many ways in which it promotes survival and well-being from the risks that occur when, in the absence of empathy and fellow-feeling, it hardens and debilitates psychological growth. As for the long-standing controversies about the respective roles of nature and nurture, brain science is helping us come closer to understanding their remarkable interactions; something that may well end by extending the scope for individual and collective initiative. In this context, however, a new technology such as brain imaging can present its own risks of premature closure, if preliminary research findings are heralded as validating vast, as yet uncorroborated, speculations about human nature.

When it comes to the pervasive belief among adherents of different religious and political persuasions that skeptics must be living in illusion, however, there is little to suggest that scientific or other evidence will soon qualify such convictions. At issue, in this regard, are mutual perceptions that those on the other side have arrived at premature closure – that they simply do not see what should be self-evident.

The second sense of the term "scope," as freedom, latitude, and leeway, invites questions about the possible limits to freedom in the pursuit of happiness. Such limits can be empirical: there is much that we might like to do to be happier that is simply not possible. Other limits are moral in nature, concerning what we owe to others and to ourselves, and what we ought to do or not do in the pursuit of happiness. Such moral limits may be built into the definition of "happiness," as was done by Aristotle and all who see virtue or character as necessary for people to be called happy; or the limits may be thought, as by Kant, to be indispensable regard-less of what form of happiness people want to pursue. For Aristotle, speaking of, for example, "a happy torturer" would be a contradiction in terms.[4] For Kant, it would be beside the point, given that far more than happiness matters for someone aiming to lead a worthwhile life, including respect for oneself as a moral agent and for the fundamental rights of others. In either case, the moral limits imposed on the pursuit of happiness are central.

I have argued for the greatest possible freedom and leeway in the pursuit of happiness, subject to such moral limits. There is no one view of happiness that should exclude all others, much less be imposed on the recalcitrant. But the pursuit cannot merely involve "choosing happiness" as many advice manuals propose. Pursuits of happiness that abide by fundamental moral values differ crucially from those that call for deceit, violence, betrayal.

Misguided hopes for future happiness still play a central role in fanning many of the most debilitating practices of violence and fraud and exploitation world-wide – not only the happiness that might come from boundless wealth and power but also that of serving a patriotic or religious cause, however brutal. For example, it is possible that the al-Qaeda pilots who guided the planes into the World Trade Center Towers experienced some cataclysmic high, some blend of bliss, exultation, glory, and blinding power in the moments just before the explosive impact. We cannot know, and surely terror may have been part of the blend as well; but judging from the testimony of individuals who have attempted to carry out suicide bombings and survived, at least some of the pilots may well have experienced intense happiness. What we can know is that the pilots had been *promised* everlasting happiness in a future paradise, and that some of their parents, siblings, sometimes children, express happiness at their actions and certainty that they have achieved lasting bliss.

Such death-dealing conceptions of happiness shed unexpected light on ordinarily benign claims such as Willa Cather's "That is happiness: to be dissolved into something complete and great," or on injunctions such as Joseph Campbell's to "Follow your bliss."[5]

In this harsher light, Jonathan Swift's caustic definition of happiness or "Felicity," mentioned above, takes on new meaning: "The Possession of being well Deceived; The Serene Peaceful State of being a Fool among Knaves."[6] Although far from all who propound doctrines of happiness or propose paths to achieving it are "knaves" in Swift's sense, many who do so have political aims, no matter how otherworldly their professed doctrine. His words remind us to look with caution at the conflicts that visions of happiness can inspire and at the starkly different life choices that they may require.

In earlier books, especially *Lying*, *Secrets*, *Mayhem*, and *Common Values*, I have focused on fundamental moral values, such as curbs on deceit, violence, and betrayal, and basic forms of reciprocity and nurturing that are recognized in every society.[7] But while these values are shared in the sense that no society or even family could survive without them, they are not shared in another sense – that of being recognized by all as applying to outsiders, strangers, or enemies. Nevertheless, the very fact that these values have been found necessary in all societies offers a basis for dialogue about how to extend them; for rejecting, in their name, practices such as slavery, torture, and genocide, and for critiques of states, groups, and doctrines that endorse such practices.

As I have explored the discordant claims about what true or real happiness is, I have come to find a closer connection to those earlier books than I anticipated at first. Examples abound of how accepting a particular vision of happiness can lead to practical choices that either adhere to or violate fundamental moral values; and of how such choices can ennoble or degrade those who make them.

* * * *

Looking back to when I first set out to explore the subject of happiness, I find that the undertaking has been just the kind of adventure I hoped for. I have sought to shift between paying close attention to particular depictions of happiness and examining them together, alternately glowing and shadowy, brilliantly colored or vanishingly pale, exuberant or meditative, inviting all comers or excluding all but a select few, invested with aching hope or with fear. In so doing, I've come to envisage them as comparable to some gorgeous many-hued tapestry, with each separate part uniquely alive, drawing the eye into its depth. They stand out the more luminously when delineated against the dark backdrop of suffering, injustice, and death. Those who see only that backdrop – "the evils under the sun" in Ecclesiastes – regard such conceptions of happiness as illusory. Others, who succeed in blotting out the backdrop, may feel satisfied with superficial surrogates or feel-good messages. It is the vision of those who can encompass both the evils and the potential for happiness who help the rest of us to fathom what Montaigne, in his *Essays*, was one of the first to call the "human condition":

> I set forth a humble and inglorious life; that does not matter. You can tie up all moral philosophy with a common and private life just as well as with a life of richer stuff. Each man bears the entire form of the human condition.[8]

NOTES

I. Luck

1. See my *Alva Myrdal: A Daughter's Memoir* (Addison Wesley, 1991), 90, 107.
2. Desmund Tutu, *No Future without Forgiveness* (Doubleday, 1999), 5–6.
3. Ingrid Betancourt, quoted in the *Boston Globe*, July 4, 2008, A4.
4. Norman Vincent Peale, *The Power of Positive Thinking* (Prentice-Hall, 1952).
5. Horace Fletcher, *Happiness as Found in Forethought Minus Fearthought* (Stone, 1897). Howard Mumford Jones, in *The Pursuit of Happiness* (Harvard University Press, 1953), 156, cites a 1913 edition of Fletcher's book and points to the difference between such books, increasingly popular in the United States in the twentieth century, and what Thomas Jefferson and others had meant by "the pursuit of happiness."
6. Ronald F. Inglehart, Roberto Foa, Christopher Peterson, and Christian Welzel, "Development, Freedom, and Rising Happiness: A Global Perspective 1981–2007," in *Perspectives on Psychological Science*, vol. 3, July 2008, 264–85.
7. Robert Mauzi, *L'Idée du bonheur au xviiie siècle* (Librairie Armand Colin, 1960); V. J. McGill, *The Idea of Happiness* (Frederick A. Praeger, 1967); Peter Quennell, *The Pursuit of Happiness* (Constable, 1988).

II. Experience

1. Jean-Jacques Rousseau, *Les Confessions* (Paris: Editions Garnier Freres, 1964), 259: my translation. See also Jean-Jacques Rousseau, *Confessions*, ed. P. N. Furbank (Alfred A. Knopf, 1992), 206.
2. Frederick Douglass, *Bondage and Freedom*, in *Douglass Autobiographies* (The Library of America, 1994), 350.
3. Charles Darwin, *The Expression of the Emotions in Man and Animals*, 3rd edn, ed. Paul Ekman (Oxford University Press, 1998), 195.
4. Paul Ekman, ibid., "Afterword," 375–7. See also Paul Ekman and R. J. Davidson, *The Nature of Emotion* (Oxford University Press, 1994).
5. Jerome Kagan, *Three Seductive Ideas* (Harvard University Press, 1998), 173.
6. Antonio Damasio, *The Feeling of What Happens: Body and Emotion in the Making of Consciousness* (Harcourt, 1999), 9. Damasio speaks of "an autobiographical memory which is constituted by implicit memories of multiple instances of the past and of the anticipated future. . . . [It] grows continually with life experience but can be partly remodeled to reflect new experiences," 172.

7. See Carl Darling Buck, *A Dictionary of Selected Synonyms in the Principal Indo-European Languages* (University of Chicago Press, 1949), 1104–6. See also Grant Duncan, "After Happiness," *Journal of Political Ideologies*, vol. 85, 2007, 87–8; and A. Wierzbicka, " 'Happiness' in Cross-Linguistic and Cross-Cultural Perspective," *Daedalus*, vol. 133, Spring 2004, 34–43.

8. See Michael Eid and Randy L. Larsen, eds, *The Science of Subjective Well-Being* (Guilford Press, 2008); and Felicia A. Huppert, Nick Baylis, and Barry Keverne, eds, *The Science of Well-Being* (Oxford University Press, 2003). In a related approach, the Chilean biologist and neuroscientist Francisco Varela developed techniques of "neurophenomenology," starting in the late 1980s to investigate moment-by-moment shifts in perception, combining introspective first-person reports with brain imaging and other third-person techniques. See Francisco Varela, "Neurophenomenology: A Methodological Remedy to the Hard Problem," *Journal of Consciousness Studies*, vol. 3, 1996, 1313–17; and Daniel Goleman, *Destructive Emotions: How Can We Overcome Them?* (Bantam Books, 2003), 306–9, 285–6.

9. Tom Siegfried, "New Scientific Discipline Uses Brain Images to Shed Light on Social Behavior," *Dallas Morning News*, May 17, 2004, E1.

10. Stephen S. Hall, "Is Buddhism Good for Your Health?", *New York Times Magazine*, September 14, 2003, 46.

11. Owen Flanagan, *The Really Hard Problem: Meaning in a Material World* (MIT Press, 2007), 160. Flanagan has since explained that his Yes was ironic.

12. Matthieu Ricard, *Happiness* (Little, Brown, 2006), 265–6.

13. William James, *Varieties of Religious Experience*, ed. Fredson Bowers (Harvard University Press, 1985), 71.

14. Ibid., 411.

15. William James, letter to Frances Morse, quoted by Richard R. Niebuhr in "William James on Religious Experience," in Ruth Anna Putnam, ed., *The Cambridge Companion to William James* (Cambridge University Press, 1997), 215.

16. "The Memoirs of Sinuhit, an Egyptian Papyrus," tr. Flinders Petrie, *Autobiography in the Ancient World* (University Library of Autobiography, 1918), vol. I, 13.

17. P. D. James, *Time to Be in Earnest: A Fragment of an Autobiography* (Ballantine Books, 1999), 187.

18. Lionel Trilling, *Sincerity and Authenticity* (Harvard University Press, 1972), 19, 24.

19. See Arnaldo Momigliano, *The Development of Greek Biography* (Harvard University Press, 1971) on Greek and Roman biography and autobiography; and Ihor Ševčenko, "Quatre leçons de littérature byzantine," in *La Civiltà bizantina dal XII al XV secolo* ("L'Erma" di Bratschneider, 1982), 111–88.

20. See George Duby, "Solitude: Eleventh to Thirteenth Century," tr. Arthur Goldhammer, in George Duby, *A History of Private Life* (Harvard University Press, 1998), vol. II, 509, on the flourishing of portraiture and biography in the twelfth century and the stress on the autonomy of the individual; and Linda Georgianna, *The Solitary Self: Individuality in the Ancrenne Wisse* (Harvard University Press, 1981), on the focus on individualism and personal life as a hallmark of the twelfth-century Renaissance.

21. Edith Hamilton, *The Greek Way* (The Modern Library, 1942), 30.

22. Michel de Montaigne, "To the Reader," in *The Complete Essays of Montaigne*, tr. Donald Frame (Stanford University Press, 1965), 2; Queen Christina, *Självbiografi och aforismer* (Natur och Kultur, 1959), 7–8. For a discussion of

these and other autobiographical works, see my chapter "Autobiography as Moral Battleground" in Daniel Schacter and Elaine Scarry, eds, *Memory, Brain, and Belief* (Harvard University Press, 2000), 305–24.

23. Philippe Lejeune, *Le Pacte autobiographique* (Seuil, 1975), 9.

24. *The Life of Teresa of Jesus: The Autobiography of St Teresa of Avila*, tr. E. Allison Peers (Doubleday, 1960), 172, 174.

25. Henry David Thoreau, *Journal*, July 16, 1851, in Walter Harding, ed. *Henry David Thoreau: Selections from the Journals* (Dover Publications, 1995), 8.

26. Vladimir Nabokov, *Speak, Memory* (G. P. Putnam's, 1966), 139.

27. *The Autobiography of Charles Darwin, 1809–1882*, ed. Nora Barlow (W.W. Norton, 1958), 81.

28. Claire Pic, journal entry, February 24, 1867, in Philippe Lejeune, *Le Moi des demoiselles: Enquête sur le journal de jeune fille* (Editions du Seuil 1993), 248. [My translation.]

29. Ibid., See also Lejeune, *Le Pacte autobiographique*.

30. Iris Murdoch, *Metaphysics as a Guide to Morals* (Penguin Press, 1992), 337.

31. Iris Murdoch, *The Sovereignty of Good* (Routledge & Kegan Paul, 1985), 87.

32. Robert Fulford, "My Church: The Mind's 'Theatre of Simultaneous Possibilities'," *National Post*, December 22, 2007, quoting William James's *Principles of Psychology*.

33. Robert Nozick, *Anarchy, State, and Utopia* (Basic Books, 1974), 42.

34. Ibid., 43–4.

35. Ed Diener and Robert Biswas-Diener mention having used the example with thousands of students and finding that 95 percent "opt for work, social connections, and pain over orgasmic bliss," in *Happiness: Unlocking the Mysteries of Psychological Wealth* (Blackwell, 2008), 241.

36. Robert Nozick, "Happiness," in *The Examined Life: Philosophical Meditations* (Simon & Schuster, 1989), 99–117.

37. Ibid., 105.

38. Ibid., 117.

39. Ibid., 105–6.

40. Montaigne, "To the Reader," in *Essays*, 2.

41. I discuss a third perspective, that of the relatives and others mentioned in autobiographical accounts who have a personal stake in what is being said about them, in "Autobiography as Moral Battleground," 307–24.

42. Montaigne, *Essays*, 101.

43. Montaigne, *Essays*, 721.

44. Ibid., 838.

45. Ibid., 798.

46. Ibid., 156.

47. Ibid., 850.

48. Ibid., 850–1.

49. Stanley Cavell, *Pursuits of Happiness: The Hollywood Comedy of Remarriage* (Harvard University Press, 1981), 238.

50. Among thinkers I discuss in this book, Blaise Pascal, contemptuous though he could be of Montaigne, nevertheless read him with sufficient care to lift most of his Latin quotations from him. The young Immanuel Kant emulated Montaigne in his *Observations on the Beautiful and the Sublime*, strolling from topic to topic and throwing off cascades of personal impressions and bits of information (and misinformation) gathered from his readings. Rousseau declared, in his last

book, the *Reveries of the Solitary Walker*, that his self-examination would be like Montaigne's in everything, save that Montaigne was addressing his readers, Rousseau only his own self. And in his essay "Montaigne; or the Skeptic", Ralph Waldo Emerson, recalling how he came across Montaigne's *Essays* in his father's library, told of "the delight and wonder in which I lived with it. It seemed to me as if I had myself written the book, in some former life, so sincerely it spoke to me of my thought and experience." See William Gilman, ed., *Selected Writings of Ralph Waldo Emerson* (Signet, 2003) 398.

51. Virginia Woolf, "Montaigne," in *The Common Reader*, First Series (Harvest Books, 1984), 67.

52. See my entry on the positive and negative formulations of the Golden Rule in *The Oxford Companion to Philosophy*, ed. Ted Honderich (Oxford University Press, 1995), 321, and my book *Common Values* (University of Missouri Press, 2002), 14–15.

53. David Hume, "My Own Life", in *Essays Moral, Political and Literary* (Henry Frowte, 1904) 615.

54. Ralph Waldo Emerson, "Nature." The same passage ends less vividly with "Almost I fear to think how glad I am," in Robert E. Spiller and Alfred R. Ferguson, eds, *Nature, Addresses, and Lectures* (Harvard University Press, 1975), "Nature," 10.

55. Virginia Woolf, "A Sketch of the Past," in *Moments of Being*, ed. Jeanne Schulkind (Harcourt, Brace, 1985), 64–5.

56. Ibid., 67.

57. François, duc de La Rochefoucauld, *Maxims*, tr. Constantine FitzGibbon (Allen Wingate, 1957), 52.

58. Ivor A. Richards, *Principles of Literary Criticism* (Kegan, Paul, Trench, Trübner, 1925), 40.

III. Discordant Definitions

1. Alexander Pope, *An Essay on Man*, in *The Poems of Alexander Pope*, ed. John Butt (Yale University Press, 1963), Epistle IV, 536.

2. Ibid., 536–7.

3. Ibid., 564.

4. V. J. McGill, *The Idea of Happiness* (Frederick A. Praeger, 1967), 5–6. McGill adds two provisos in stating the root meaning: "that the satisfied desires can include satisfactions that are not *preceded* by specific desires for them, but come by surprise"; and that "the most favorable ratio" should read "the most favorable ratio or some approximation to it" so as to go along with the common judgment that one man can be happier than another.

5. Seneca, *Epistulae Morales*, tr. Richard M. Gummere (Loeb Classical Library, Harvard University Press, 1967); for example, vol. I, Epistles V, XIII, 23–5, 73–83.

6. Charles Fourier, *Théorie des quatre mouvements et des destinées générales*; see Charles Fourier, *Theory of the Four Movements*, ed., Gareth Steadman Jones and Ian Patterson (Cambridge University Press, 1996), 95.

7. McGill, *Idea of Happiness*, "Eudaemonism vs. Hedonism," 262–317. See also Richard M. Ryan and Edward L. Deci, "On Happiness and Human Potentials: A Review of Research on Hedonic and Eudaimonic Well-Being," *Annual Review of Psychology*, vol. 52, 2001, 141–66.

8. McGill, *Idea of Happiness*, 5. McGill notes that in arguing that happiness had to be an activity, Aristotle anticipated and took issue with the Stoics, Plotinus, and Augustine, for whom happiness (or perfect happiness) was a state of the soul. And in rejecting the view that the possession of virtue *alone* constituted felicity, Aristotle took express issue with the Platonic Socrates and the teachings of the Stoics. There were two positions, McGill adds, that Aristotle did not anticipate: the utilitarian formula of "the greatest pleasure of the greatest number," with its radical egalitarian implications; and Kant's rejection of happiness as the supreme good and his putting in its place doing one's duty, or being worthy of happiness.

9. Aristotle, *Nicomachean Ethics*, tr. Terence Irwin (Hackett, 1985), 1098a12–21, 17. See also Julia Annas, *The Morality of Happiness* (Cambridge University Press, 1993), 142–58; Kwame Anthony Appiah, *Experimental Ethics* (Harvard University Press, 2008), ch. 5; Sarah Broadie, *Ethics with Aristotle* (Oxford University Press, 1991), ch. 1; Sarah Broadie, *Aristotle and Beyond: Essays on Metaphysics and Ethics* (Cambridge University Press, 2007); Monique Canto-Sperber, "Bonheur," in Monique Canto-Sperber, ed., *Dictionnaire de l'ethique et de la philosophie morale*, 4th edn (Presses Universitaires de France, 2004), vol. iv, 201; John M. Cooper, *Reason and Human Good in Aristotle* (Harvard University Press, 1995), 89–100; Martha Nussbaum, *The Fragility of Goodness: Luck and Ethics in Greek Tragedy and Philosophy* (Cambridge University Press, 1986); and Bernard Williams, *Ethics and the Limits of Philosophy* (Harvard University Press, 1985), 34.

10. Broadie, *Aristotle and Beyond*, 166.

11. Thomas Merton, *No Man Is an Island* (Walker, 1955), 175; Ayn Rand, *The Virtue of Selfishness: A New Concept of Egoism* (Signet Books, 1964), 31.

12. Richard Layard, *Happiness: Lessons from a New Science* (Penguin Books, 2005), 12.

13. Ed Diener, "Subjective Well-Being: The Science of Happiness and a Proposal for a National Index," *American Psychologist*, January 2000, 34. See also Ed Diener, E. M. Suh, Richard E. Lucas, and Heidi L. Smith, "Subjective Well-Being: Three Decades of Progress," *Psychological Bulletin*, vol. 125, 1999, 277; Richard Easterlin, "Explaining Happiness," paper presented to the National Academy of Sciences, May 23, 2003; Sonja Lyubomirsky, *The How of Happiness: A Scientific Approach to Getting the Life You Want* (Penguin Press, 2008); Martin P. Seligman, *Authentic Happiness: Using the New Positive Psychology to Realize Your Potential for Lasting Fulfillment* (Free Press, 2002) 261; Ruut Veenhoven, Happiness as an Aim in Public Policy," in Alex Linley and Stephen Joseph, eds, *Positive Psychology in Practice* (John Wiley, 2004), 6.

14. Richard B. Brandt, "Happiness," in *Encyclopedia of Philosophy* (Macmillan, 1967), vol. III, 413–14.

15. Wayne Sumner, "Something in Between," in Roger Crisp and Brad Hooker, eds, *Well-Being and Morality: Essays in Honour of James Griffin* (Oxford University Press, 2000), 15.

16. See Sarah Broadie's compact formula of Aristotle's view, above, note 10.

17. Derek Parfit, *Reasons and Persons* (Clarendon Press, 1984), 4, 499–502. See also T.M. Scanlon, "Value, Desire and Quality of Life," in Martha Nussbaum and Amartya Sen, eds, *The Quality of Life* (Oxford University Press, 1999), 185–200; and Sissela Bok, "Comment," in ibid., 201–7; and Wayne Sumner, "Objective Theories," in *Welfare, Happiness and Ethics* (Oxford University Press, 1996).

18. Martha Nussbaum, "Non-Relative Virtues: An Aristotelian Approach," in Nussbaum and Sen, eds, *The Quality of Life*, 259–60.

19. John Kekes, "Happiness," in Lawrence C. Becker and Charlotte B. Becker, eds, *Encyclopedia of Ethics*, 2nd edn (Routledge, 2001), vol. II, 650. See also James. P. Griffin, "Happiness," in Edward Craig, ed., *The Routledge Encyclopedia of Philosophy* (Routledge, 1998), 226–9; and Richard J. Norman, "Happiness," in Ted Honderich, ed., *Oxford Companion to Philosophy* (Oxford University Press, 1995), 332–3.

20. See Sumner, "Something in Between," 15–17.

21. Robert Nozick, *The Examined Life: Philosophical Meditations* (Simon & Schuster, 1989), 111.

22. Compare Aristotle's view: "Everyone who has the power to live according to his own choice should . . . set up for himself some object for the good life to aim at . . . with reference to which he will then do all his acts, since not to have one's life organized in view of some end is a sign of great folly," *Eudemian Ethics*, tr. J. Soloman, in Jonathan Barnes, ed., *The Complete Works of Aristotle* (Princeton University Press, 1984), vol. ii, 1923 (1214b6–14). As Cooper points out in *Reason and Human Good*, 94, Aristotle held that while everyone ought to have their desires organized in such a way, not everyone does. See Elizabeth Telfer, *Happiness* (St Martin's Press, 1980), 2–5. Robert Nozick, in *Anarchy, State, and Utopia*, (Basic Books, 1974) 50, makes a different claim: a person's "shaping his life in accordance with some overall plan is his way of giving meaning to his life . . ." John Kekes indicates that looking both forwards and backwards at one's life matters: "I want to consider a man judging his life in midstream": "Happiness," *Mind*, vol. 91, 1982, 370. For Kekes, too, having a life-plan matters, though he specifies that much can go wrong with that life-plan, by error or through self-deception or other misjudgment – something that is often only evident to outsiders. See also Thomas W. Pogge, "Human Flourishing and Universal Justice," in E. Paul, Fred D. Miller, Jr., and Jeffrey Paul, eds, *Human Flourishing* (Cambridge University Press, 1999).

23. John Rawls, *A Theory of Justice*, revised edition (Harvard University Press, 1999), 359. See also 480–2.

24. Michel de Montaigne, *The Complete Essays of Montaigne*, tr. Donald Frame (Stanford University Press, 1965), 243.

25. Alison Jaggar, in "Feminism and Moral Philosophy," *APA Newsletter on Feminism and Philosophy*, Spring 2000, 200–5, points to critiques of Rawls and others as idealizing the capacities and "individualist planning ethics" of middle- or upper-class people in capitalist societies.

26. Aristotle, *Nicomachean Ethics*, 1101a15–18, 26–7.

27. Ibid., 1100a69. See Martha Nussbaum, "Mill between Aristotle and Bentham," *Daedalus*, Spring 2004, 61.

28. *Herodotus*, tr. A. D. Godley (Loeb Classical Library, Harvard University Press, 1975), vol. I, 34–5. For a discussion of Solon's replies to Croesus, see Darrin M. McMahon, *Happiness: A History* (Atlantic Monthly Press, 2006), 1–7; and Helen Small, *The Long Life* (Oxford University Press, 2008), 53–61.

29. Julia Annas, "Happiness as Achievement," *Daedalus*, Spring 2004, 45. See, more generally, Julia Annas, *The Morality of Happiness* (Oxford University Press, 1993).

30. *Herodotus*, 38–9.

31. Sigmund Freud, *Civilization and Its Discontents*, tr. and ed. James Strachey, with an Introduction by Louis Menand and a Biographical Afterword by Peter Gay (W. W. Norton, 2005), 53.

32. Jonathan Lear, *Happiness, Death, and the Remainder of Life* (Harvard University Press, 2000), 129.

33. Daniel Kahneman, "Objective Happiness," in Daniel Kahneman, Ed Diener, and Norbert Schwarz, eds, *Well-Being: The Foundations of Hedonic Psychology* (Russell Sage Foundation, 1999), 3–25.

34. See the Gallup World Poll, at www.gallup.com/consulting/worldpoll; and Ronald F. Inglehart, Roberto Foa, Christopher Peterson, and Christian Welzel, "Development, Freedom, and Rising Happiness: A Global Perspective," in *Perspectives on Psychological Science*, vol. 3, July 2008, 264–85. See also the World Values Survey, at www.worldvaluessurvey.org; and discussion of these and other studies in Ed Diener, Richard E. Lucas, Ulrich Schimmack, and John Helliwell, *Well-Being for Public Policy* (Oxford University Press, 2009).

35. Seneca, "On the Happy Life," in Seneca, *Moral Essays II*, tr. John W. Basore (Loeb Classical Library, Harvard University Press, 1990), 140–1: "Therefore true happiness is founded upon virtue." ("Ergo in virtute posita est vera felicitas.")

36. Cicero, *Tusculan Disputations*, tr. J. E. King (Loeb Classical Library, Harvard University Press, 1971), vol. V, 503, 105. See also Seneca, *Epistulae Morales*, vol. ii, lxvi, 15, quoting Epicurus.

37. Diogenes Laertius, quoting Plato, in *Lives of Eminent Philosophers*, tr. R. D. Hicks (Loeb Classical Library, Harvard University Press, 1972), vol. I, 345.

38. John M. Cooper, in "Aristotle on the Goods of Fortune," *Philosophical Review*, vol. 94, April 1985, 173–96, elucidates what Aristotle means by "goods of fortune," as relating to *eudaimonia*. A narrower view contrasts these goods with two other categories: goods of the soul (such as virtues and pleasure) and goods of the body. In a broader view, "the only goods that are not external are the virtues of intellect and character themselves, innate endowments of mind and personality such as cheerfulness or a good memory, pleasure, and knowledge."

39. See Martha Nussbaum, *The Therapy of Desire: Theory and Practice in Hellenistic Ethics* (Princeton University Press, 1994).

40. Immanuel Kant, *Groundwork of the Metaphysics of Morals* (1785) in Mary Gregor, tr. and ed., *Immanuel Kant: Practical Philosophy* (Cambridge University Press, 1996), 49.

41. Ibid., 50.

42. Immanuel Kant quoting Juvenal, in *The Critique of Practical Reason* (1788), in Gregor, tr. and ed., *Immanuel Kant*, 267.

43. Ibid., 156. In the *Critique of Judgment*, Kant's definition is if possible even more uncompromising: "happiness is the satisfaction of all our desires; *extensive*, in regard to their multiplicity; *intensive*, in regard to their degree; *protensive*, in regard to their duration." This is the same definition that he had already set forth a decade earlier (1781) in his *Critique of Pure Reason*: tr. Norman Kemp Smith, *Immanuel Kant's Critique of Pure Reason* (St Martin's Press, 1961), 636.

44. Friedrich Hegel, *The Phenomenology of Mind*, cited by McGill, *Idea of Happiness*, 110; John Rawls, *Lectures on the History of Moral Philosophy*, ed. Barbara Herman (Harvard University Press, 1999), 316–17, Griffin, "Happiness," 227; Christine Korsgaard, "Just Like All the Other Animals of the Earth," *Harvard Divinity Bulletin*, Autumn 2008, 59. For discussions of Kantianism and Kant, see Annas, *The Morality of Happiness*, 448–52; Barbara Herman, *Moral Literacy* (Harvard University Press, 2007), 178–83; Raymond Angelo Belliotti, *Happiness is Overrated* (Rowman & Littlefield, 2004), 45–6; Stephen Engstrom and Jennifer Whiting, eds, *Aristotle, Kant, and the Stoics: Rethinking Happiness and Duty* (Cambridge University Press, 1996); and Darrin M. McMahon, *Happiness: A History* (Atlantic Monthly Press, 2006), 212–15.

45. Immanuel Kant, *Lectures on Ethics*, tr. Louis Infield (Harper & Row, 1963), 67.

46. Immanuel Kant, *Anthropology from a Practical Point of View*, tr. Victor Lyle Dowdell (Southern Illinois University Press, 1978), 206–7.

47. Bernard Williams, *Ethics and the Limits of Philosophy* (Harvard University Press, 1985), 46.

48. See Peggy A. Thoits and Lyndi N. Herwitt, "Volunteering Work and Well-Being," *Journal of Health and Social Behavior*, vol. 47, 2001, 215; Seligman, *Authentic Happiness*, 10, 42–3; Matthieu Ricard, *Happiness* (Little, Brown, 2006), ch. 17; Shankar Vedamtan, "If It Feels Good To Be Good, It Might Be Only Natural," *Washington Post*, May 28, 2007, A01.

49. Stephen Post, "Altruism, Happiness, and Health: It's Good To Be Good," *International Journal of Behavioral Medicine*, vol. 12, no. 2, 2005, 66–77.

50. Lyubomirsky, *The How of Happiness*, 125–35.

51. Jorge Moll, Frank Krueger, Roland Zahn, Matteo Pardini, Ricardo de Olivera-Souza, and Jordan Grafman, "Human fronto-mesolimbic networks guide decisions about charitable donation", *Proceedings of the National Academy of Sciences of the United States of America*, vol. 103, no. 42, October 17, 2006, 15623–28. See also Patricia Smith Churchland, "The Impact of Neuroscience on Philosophy", *Neuron*, vol. 60, November 6, 2008, 409–11.

52. Denis Diderot, cited by G. Mangeot in "Les Réflexions sur le bonheur de la Marquise du Châtelet," in *Mélanges Lanson* (Librairie Hachette 1922), 277.

53. Friedrich Nietzsche, *The Anti-Christ*, par. 2, in *Twilight of the Idols/ The Anti-Christ*, tr. R. J. Hollingdale (Penguin Books, 1990), 127–8.

54. Ellen Key, *Barnets århundrande* (The Century of the Child), [my translation] (Albert Bonniers Förlag, 1912), 63.

55. W. B. Yeats, diary entry, quoted in Jay Parini, *Why Poetry Matters* (Yale University Press, 2008), 51–2.

56. Nozick, *The Examined Life*, 117.

57. Stanley Cavell, *Pursuits of Happiness: The Hollywood Comedy of Remarriage* (Harvard University Press, 1981), 45. See Mark Kingwell, *In Pursuit of Happiness* (Crown Publishers, 1998), 326, for a discussion of Cavell's view.

58. Roy Strong, *On Happiness* (Long Barn Books, 1997), 47–8.

59. See John Searle, *Speech Acts* (Cambridge University Press, 1970), 55; and my discussion, with respect to defining "violence," in *Mayhem: Violence and Public Entertainment* (Perseus Books, 1999), 6–7.

60. See Charles Stevenson, *Ethics and Language* (Yale University Press, 1950), 210; and Sören Halldén, *True Love, True Humour, and True Religion* (C. W. Gleerup, 1960).

61. Richard Robinson, *Definition* (Clarendon Press, 1950).

62. Jonathan Swift, "A Digression on Madness," in A. C. Guthkelch and D. Nichol Smith, eds, *A Tale of a Tub* (Clarendon Press, 1958), 171.

63. Iris Murdoch, *The Nice and the Good* (Penguin Books, 1978), ch. 22; Lyubomirsky, *The How of Happiness*, 32.

IV. "On the Happy Life"

1. Seneca, "On the Happy Life," in *Moral Essays* II, tr. John W. Basore (Loeb Classical Library, Harvard University Press, 1932), 107. I have made minor changes in the translation to come closer to Seneca's text.

2. See Miriam Griffin, *Seneca: A Philosopher in Politics* (Oxford University Press, 1992), 6–19.
3. Seneca, "Happy Life", 99.
4. Ibid., 145.
5. Ibid., 99.
6. Ibid., 107.
7. Ibid., 157. The "indifferents," or *indifferentia*, mentioned by Seneca, represented the Stoic *adiaphora*, a technical term characterizing the things that lie outside of the categories of *virtus* and *decedus*, the sole good and the sole evil.
8. Ibid.
9. Ed Diener and Robert Biswas-Diener, *Happiness: Unlocking the Mysteries of Psychological Wealth* (Blackwell, 2008), 91, 110. Because the question of which way causation runs is difficult, when determining whether happier people tend to be wealthier (or healthier, etc.) or the reverse, research on lotteries is especially informative, since lottery winners are usually no different in this regard than those who do not win.
10. Ibid., 111.
11. Ibid. See also Ed Diener, E. M. Suh, Richard E. Lucas, and Heidi L. Smith, "Subjective Well-Being: Three Decades of Progress," *Psychological Bulletin*, vol. 125, 1999, 289.
12. Seneca, *De beneficiis*, in *Seneca: Moral Essays*, III, tr. John W. Basore (Loeb Classical Library, Harvard University Press, 1964); see also Seneca's more succinct Epistle 81 to Lucilius, "On Benefits," in Seneca, *Epistulae Morales*, tr. R. M. Gummere (Loeb Classical Library, Harvard University Press, 1962), vol. ii 219–41. For discussion of Seneca's views in the context of Stoicism, see Brad Inwood, *Reading Seneca: Stoic Philosophy at Rome* (Clarendon Press, 2005).
13. Seneca, "To Marcia on Consolation," in *Moral Essays*, II, tr. John W. Basore (Loeb Classical Library, Harvard University Press, 1932), 3–97. I discuss this essay in "Rethinking Common Values," in Ken Booth and Tim Dunne, eds, *Worlds in Collision: Terror and the Future of Global Order* (London: Palgrave Macmillan, 2002).
14. Seneca, "To Marcia," 63.
15. Seneca, *Epistulae Morales*, Letter XXVII, 197.
16. Arthur Darby Nock, *Conversion* (Oxford University Press, 1963), 7, 165.
17. Augustine, "The Happy Life," tr. Ludwig Schopp, in *Writings of St Augustine*, vol. I (CIMA, 1948), 41–84, at 43–4. Both Seneca's and Augustine's essays have been translated into French by Marc Vaillancourt and published together in *Le Bonheur: De Vita Beata; De Beata Vita* (42e Parallèle, 2005). The Theodore to whom Augustine dedicated his essay was Mallius Theodorus, with whom he had studied Neoplatonic texts in Milan. Later in life, Augustine regretted his praise of Theodore, whom he criticized for having reverted to paganism.
18. Homer, *The Odyssey*, tr. Robert Fitzgerald (Alfred A. Knopf, 1989), Book 8, 261–3. See Gregory Nagy, *Poetry as Performance: Homer and Beyond* (Cambridge University Press, 1996), 187–206, for a discussion of the *Iliad* and the *Odyssey* as scripture for the Greeks, in the sense of both sacred and known by all. See also Michael O'Loughlin, *The Garlands of Repose: The Literary Celebration of Civic and Retired Leisure* (University of Chicago Press, 1978), 11–31.
19. For Augustine's stay at Cassiciacum, see Peter Brown, *Augustine of Hippo* (University of California Press, 1975), 101–28; Michael P. Foley, "The Other

Happy Life: The Political Dimensions to St Augustine's Cassiciacum Dialogues," *Review of Politics*, vol. 65, Spring 2003, 165–84; Paula Fredriksen, *Augustine and the Jews* (Doubleday, 2008), 122–34; and Garry Wills, *Saint Augustine* (Viking, 1999), 48–56.

20. Augustine, "The Happy Life," 50–1.
21. Ibid., 57.
22. Ibid., 59.
23. Ibid. For the evolution of Augustine's attitude toward celibacy, see Fredriksen, *Augustine and the Jews*, 126–38.
24. Augustine, "The Happy Life," 83.
25. Ibid., 84.
26. Augustine, *Retractationes Italian and Latin* (Città Nuova, 1993).
27. Boethius, *The Consolation of Philosophy* (524), tr. Richard H. Green (Bobbs-Merrill, 1962), 8. See also Mark Kingwell, *Better Living: In Pursuit of Happiness from Plato to Prozac* (Viking, 1998), 255–61.
28. Thomas Aquinas, *Summa Theologica*, tr. Fathers of the English Dominican Province (Christian Classics, 1981) vol. II, First Part of the Second Part, Questions 1–5, 583–615.
29. Ibid., Question 3, article 8, 601.
30. Ibid., Question 5, article 3, 610.
31. René Descartes, Letter to Elisabeth, July 21, 1645, in *Descartes: Oeuvres et Lettres* (Gallimard, 1953), 1191. For the letters of Elisabeth of Bohemia and René Descartes, see also volumes IV and V of C. Adam and P. Tannery, eds, *Oeuvres de Descartes* (J. Vrin, 1972). The translations from the French are my own.
32. For accounts of the correspondence and friendship between Elisabeth and Descartes, see Margaret Atherton, ed., *Women Philosophers of the Early Period* (Hackett, 1994), 9–21; John J. Blom, *Descartes: His Moral Philosophy and Psychology* (Harvester, 1978); Stephen Gaukroger, *Descartes: An Intellectual Biography* (Clarendon Press, 1995), ch. 10; A. C. Grayling, *Descartes: The Life and Times of a Genius* (Walker, 2007), ch. 10; Andrea Nye, *The Princess and the Philosopher: Letters of Elisabeth of the Paladine to René Descartes* (Rowman & Littlefield, 1999); Andrea Nye, "Commentary: Princess Elisabeth and Descartes: A Philosophical Correspondence," in Karen J. Warren, ed., *An Unconventional History of Western Philosophy: Conversations between Men and Women Philosophers* (Rowman & Littlefield, 2009), 177–88; and Richard Watson, *Cogito Ergo Sum: The Life of René Descartes* (Godine, 2002), ch. 10.
33. René Descartes, *Les Principes de la philosophie*, in *Descartes: Oeuvres et lettres* (Bibliothèque de la Pléiade, 1953), 555–6; and *Les Passions de l'âme*, ibid., 696–802.
34. Descartes claimed, in his letter of August 4, 1645, to have formulated these rules in his *Discourse on Method*, but readers will be hard put to discern the resemblance between the three rules and the "provisional moral code consisting of just three or four maxims" in Part Three of the *Discourse*. For a comparison, see both passages in J. B. Schneewind, ed., *Moral Philosophy from Montaigne to Kant* (Cambridge University Press, 1990), vol. I, 218–22, 225–6. Schneewind suggests that the most notable omission in Descartes's system is that he "did not develop the hints he dropped here and there about morality," and cites a letter to Chanut indicating that one reason he had not written down his thoughts about morality was that "there is no subject from which malicious people can more easily draw pretexts for slandering one," 216–17.

35. Descartes, Letter to Elisabeth, August 4, 1645 in *Oeuvres et lettres*, 1192–5. Descartes continued to discuss Seneca two weeks later, analysing his views in greater detail. He now offered a sustained critique of the notion of doing what is "according to nature": the obscurity of that concept made it hard to understand how it was supposed to relate to happiness or contentment. When Seneca finally made a claim that he could agree with, that "he is blessed who neither desires nor fears, thanks to reason," Descartes found it of little help in the absence of reasons why we ought not to fear or desire anything.

36. Descartes, Letter to Elisabeth, September 1, 1645 in *Oeuvres et lettres*, 1200–4.

37. Blaise Pascal, *Pensées*, tr. A. J. Krailsheimer (Penguin Classics, 1995), 123.

38. Ibid., 125.

39. Blaise Pascal, "Prière pour demander à Dieu le bon usage des maladies" (Prayer to ask of God the Good Use of Illnesses"), in Pascal, *Oeuvres Complètes* (Bibliothèque de la Pléiade, 1954), 605–14.

40. Bertrand Russell, *The Conquest of Happiness* (W. W. Norton, 1996), 120.

41. Pascal, *Pensées*, 37.

42. Excerpted from Blaise Pascal's notation from the night of November 23, 1654, "Mémorial," *Pensées*, 285.

43. Voltaire, "Sur les Pensées de M. Pascal," *Lettres Philosophiques* in *Mélanges* (Bibliothèque de la Pléiade, 1961), 104–53, at 104.

44. Ibid., 109.

45. Denis Diderot, *Additions aux Pensées philosophiques*, in Diderot, *Oeuvres Philosophiques* (Classiques Garnier, 1990), viii, 59.

46. Ibid., lix, 68.

47. Denis Diderot, ed., *Encyclopédie, ou dictionnaire raisonné des sciences, des arts, et des métiers*, 2nd edn (Vincent Giutini, 1758), vol. II, 272–3, entry on "Bonheur." Another entry, "Béatitude, Bonheur, Félicité," ibid., 143, distinguishes between the three terms, pointing out that "happiness marks the man rich in the goods of fortune; felicity, a man content with what he has; and beatitude the state of a man whom the immediate presence of his God fills in this world or the next." It closes, possibly tongue in cheek, by quoting the abbé Grimond: "Happiness is for the rich, felicity for the wise, beatitude for the poor in spirit."

48. Salman Rushdie, Interview with David Frost, November 26, 1993.

49. Julien Offray de La Mettrie, *De la Volupté: Anti-Sénèque ou le souverain bien; L'Ecole de la volupté, système d'Epicure*, ed. Ann Thomson (Les Editions Des Jonquières, 1996), 28. See also Robert Mauzi, *L'Idée du bonheur au xviiie siècle* (Librairie Armand Colin, 1960), 249–52; Darrin M. McMahon, *Happiness: A History* (Atlantic Monthly Press, 2006), 223–33.

50. Julien Offray de la Mettrie, *L'Homme-Machine* (1747); tr. Richard A. Watson and Maya Rybalka, *Man a Machine; and Man a Plant* (Hackett, 1994).

51. La Mettrie, *Anti-Sénèque*, 92. The Marquis de Sade commented that "the celebrated La Mettrie had been right in saying that it was necessary to wallow in filth like pigs and to find pleasure, like them, in the last degrees of corruption." See Ann Thompson, "Introduction," in *Anti-Sénèque*, 11. This is to misjudge La Mettrie, in the eyes of Thompson and most commentators.

52. La Mettrie, *Anti-Sénèque*, quoted by Mauzi in *L'Idée du bonheur au xviiie siècle*, 251.

53. Denis Diderot, *Essai sur les règnes de Claude et de Néron et sur les moeurs et les écrits de Sénèque pour servir d'instruction à la lecture de ce philosophe*, in Denis Diderot, *Oeuvres* (Editions Robert Laffont, 1994), vol. I, 971–1251. See also Douglas A. Bonneville, *Diderot's Vie de Sénèque* (University of Florida Press, 1966).

54. Diderot does not go on to say what he believed not to be true in Seneca's own case: "No virtue without happiness." For Diderot, Seneca's life had to have been unhappy in the extreme, above all during the last years when he was in Nero's employ: "the unhappiest of men, if ever virtue could be profoundly unhappy."

V. Measurement

1. Aristotle, *Nicomachean Ethics*, tr. Terence Irwin (Hackett, 1985), 1094b23–5.
2. Robert Biswas-Diener, Ed Diener, and Maya Tamir, "The Psychology of Subjective Well-being," *Daedalus*, Spring 2004, 18.
3. Jeremy Bentham, *Theory of Legislation* (Kegan, Paul, 1932), vol. I, 90, 15. Had Bentham looked at Plato's *Protagoras* (357d), however, he might have been surprised to find Socrates maintaining that, to achieve happiness, the "art of measurement" weighing pleasures and pain would be the saving principle of human life. For a discussion of Plato's views, including the passage in *Protagoras*, see Nicholas White, "Hedonism and the Measurement of Happiness," in *A Brief History of Happiness* (Oxford: Blackwell, 2006), 47.
4. Garry Wills, *Inventing America: Jefferson's Declaration of Independence* (Vintage Books, 1979), 149–61. For discussions of eighteenth-century works on happiness and its measurement see also Emma Rothschild, *Economic Sentiments: Adam Smith, Condorcet, and the Enlightenment* (Harvard University Press, 2001); and Darrin McMahon, *Happiness: A History* (Atlantic Monthly Press, 2006), 212–17.
5. Cesare Beccaria, *On Crimes and Punishments*, tr. David Young (Hackett, 1986).
6. Quoted by Mary Warnock, "Introduction," in Mary Warnock, ed., *John Stuart Mill: Utilitarianism, On Liberty, Essay on Bentham, together with Selected Writings of Jeremy Bentham and John Austin* (World Publishing, 1965), 7.
7. Jeremy Bentham, *An Introduction to the Principles of Morals and Legislation*, ed. J. H. Burns and H. L. A. Hart (Athlone Press, 1970), 12.
8. Jeremy Bentham, "Article on Utilitarianism," in Amnon Goldworth, ed., *Jeremy Bentham: Deontology, together with A Table of the Springs of Action and The Article on Utilitarianism* (Clarendon Press, 1983), 297.
9. Bentham, *Introduction to the Principles of Morals and Legislation*, 40.
10. Ibid., 38.
11. Immanuel Kant, "An Answer to the Question: 'What is Enlightenment?'," in Mary Gregor, tr. and ed., *Immanuel Kant: Practical Philosophy* (Cambridge University Press, 1996), 17.
12. Jeremy Bentham, *A Fragment on Government* (Cambridge University Press, 1988), 1.
13. Ross Harrison, "Bentham, Jeremy," in *Routledge Encyclopedia of Philosophy* (Routledge, 1988), vol. IV, 228.
14. Jeremy Bentham, *Deontology*, 130–1.
15. For discussion of the assumption among utilitarians of a social harmony of interests, according to which individuals in promoting their own interests also promote those of others, see Gunnar Myrdal, *The Political Element in the Development of Social Theory* (Routledge & Kegan Paul, 1953), ch. 2.
16. John Stuart Mill, "Bentham" (1838), in Marshall Cohen, ed., *The Philosophy of John Stuart Mill* (The Modern Library, 1961), 20 and 23. Mill had published an earlier version of the essay anonymously, in 1833, the year after Bentham died.
17. Meanwhile, Bentham was splendidly "fecund," to use his own term, in stimulating dialogues on the part of others, ever since. A fine example of such

dialogue: John Grote, *An Examination of the Utilitarian Philosophy*, ed. Joseph Bickerstadt Mayor (Thoemmes Antiquarian Books, 1990)

18. John Stuart Mill, *Autobiography*, ed. Jack Stillinger (London: Oxford University Press, 1971), 80–9. For an edition that includes the different drafts of the *Autobiography* and sections left out of the final edition, see John M. Robson and Jack Stillinger, eds, *Autobiography and Literary Essays* (University of Toronto Press, 1981). See also my review of the latter, "His Father's Children," in the *London Review of Books*, vol. 5, April 18, 1984, 20–1.

19. John Stuart Mill, *Utilitarianism*, ed. Roger Crisp, in *J. S. Mill, Utilitarianism* (Oxford University Press, 1998), footnote, 105.

20. Ibid., 13. See, for commentary, Elizabeth S. Anderson, "John Stuart Mill and Experiments in Living," *Ethics*, vol. 102, 1991, 4–26. See also Warnock, "Introduction," note 6 above; and Roger Crisp, "Editor's Introduction," in Crisp, ed., *Mill, Utilitarianism*.

21. For an analysis comparing Bentham's and Mill's approaches to measurement, see Wendy Donner, "Mill's Utilitarianism," in John Skorupski, ed., *The Cambridge Companion to Mill* (Cambridge University Press, 1998); Warnock, "Introduction," *John Stuart Mill*; and Crisp, "Editor's Introduction," *Mill, Utilitarianism*.

22. Crisp ed., *Mill, Utilitarianism*, 58.

23. John Stuart Mill, *The Subjection of Women*, ed. Susan M. Okin (Hackett, 1988), 1.

24. Ibid., 109.

25. Francis Ysidro Edgeworth, *Mathematical Psychics: An Essay on the Application of Mathematics to the Moral Sciences* (London School of Economics, 1932).

26. Ibid., 117. In a footnote, Edgeworth added that while Bentham was said to have corrected the phrase later in life, it was not corrected in his latest published works.

27. "On Hedonimetry," ibid., 99–102.

28. Ibid., 118.

29. Ibid., 77.

30. Ibid., 78.

31. Francis Y. Edgeworth, "The Hedonic Calculus," *Mind*, vol. 4, issue 15, July 1879, 404. See also *Mathematical Psychics*, 55–82.

32. *Mathematical Psychics*, 101.

33. Tom Valeo, "Use of Deep Brain Stimulation Widens," *BrainWork*, May/June 2008, 1–4.

34. Michael Eid and Randy J. Larsen, eds, *The Science of Subjective Well-Being* (Guilford Press, 2008), 113–15.

35. Antonio Damasio, *Looking for Spinoza* (Harcourt, 2003), 3.

36. See Kotaro Suzumura, "Introduction," in Kenneth J. Arrow, Amartya Sen, and Kotaro Suzumura, eds, *Handbook of Social Choice and Welfare* (Elsevier, 2002), 1–32.

37. See Richard A. Easterlin, "Subjective Well-Being and Economic Analysis: A Brief Introduction," *Journal of Economic Behavior and Organization*, vol. 45 (2001), 225–6.

38. See Randy Larsen and Zvezdana Prizmic, "Regulation of Emotional Well-Being: Overcoming the Hedonic Treadmill," in Eid and Larsen, eds, *Science of Subjective Well-Being*, 258.

39. Warner Wilson, "Correlates of Avowed Happiness," *Psychological Bulletin*, vol. 67, 294–306, quoted and reassessed in Ed Diener, E. M. Suh, Richard E. Lucas, and Heidi L. Smith, "Subjective Well-Being: Three Decades of Progress," *Psychological Bulletin*, vol. 125, 1999, 276–302.

40. Angus Campbell, Philip E. Converse, and Willard L. Rodgers, *The Quality of American Life: Perceptions, Evaluations, and Satisfactions* (Russell Sage, 1976).

41. For bibliographies, see http//worlddatabaseofhappiness.eur.nl; www.worldvaluessurvey.org; and the *Journal of Happiness Studies*.

42. Ed Diener and Robert Biswas-Diener, *Happiness: Unlocking the Mysteries of Psychological Wealth* (Blackwell, 2008), 234–43.

43. Ruut Veenhoven, "Happiness as an Aim in Public Policy," in Alex Linley and Stephen Joseph, eds, *Positive Psychology in Practice* (John Wiley, 2004), 6.

44. Daniel Kahneman, Alan B. Krueger, David Schkade, Norbert Schwarz, and Arthur Stone, "Toward National Well-Being Accounts," *American Economic Association Papers and Proceedings*, May 2004, 429–34.

45. Christoph Riedweg, citing Porphyry's Life of Pythagoras, in *Pythagoras: His Life, Teaching, and Influence* (Cornell University Press, 1955), 33.

46. Crisp, ed., *Mill, Utilitarianism*, 62. Mill's claim is hard to reconcile with his statement earlier in the same chapter, which holds that a state of happiness "is even now the lot of many, during some considerable portion of their lives."

47. Shigehiro Oishi, Ed Diener, Richard E. Lucas, and E. Suh, "Cross Cultural Variations in Predictions of Life Satisfaction," *Personality and Social Psychology Bulletin*, 1999, vol. 25, 980–90; John F. Helliwell and Robert D. Putnam, "The Social Context of Well-Being," in Felicia A. Huppert, Nick Baylis, and Barry Keverne, eds, *The Science of Well-Being* (Oxford University Press, 2007), 434–59.

48. Diener, Suh, and Smith, eds, "Subjective Well-Being," 289; Abbott L. Ferris, "Religion and the Quality of Life," *Journal of Happiness Studies*, vol. 3, 2002, 199–215; David G. Myers, "Religion and Human Flourishing," in Eid and Larsen, eds, *Science of Subjective Well-Being*, 323–43; Liesbeth Snoep, "Religiousness and Happiness in Three Nations: A Research Note," *Journal of Happiness Studies*, published online February 6, 2007; Phil Zuckerman, "The Virtues of Godlessness: The Least Religious Nations Are Also the Most Healthy and Successful," *Chronicle Review*, January 30, 2009, B4.

49. Jean-Baptiste Boyer, Marquis d'Argens, "Sur la Vie heureuse" ("On the Happy Life"), quoted in Robert Mauzi, *L'Idée du bonheur au XVIIIe siècle* (Genève, Slatkine Reprints, 1979), 110.

50. See studies cited in Sonja Lyubomirsky, *The How of Happiness: A Scientific Approach to Getting the Life You Want* (Penguin Press, 2008), 2, 39–40.

51. See Ed Diener, Christie Napa Scollon, and Richard E. Lucas, "The Evolving Concept of Subjective Well-Being: The Multifaceted Nature of Happiness," in *Advances in Cell Biology and Gerontology*, vol. 15, 2003, 187–219.

52. Ed Diener and Michael Eid, "Take-Home Messages from the Editors," in Michael Eid and Ed Diener, *Handbook of Multimethod Measurement in Psychology* (American Psychological Association, 2006), 457–63.

53. Daniel Gilbert, *Stumbling on Happiness* (Alfred A. Knopf, 2006), 64.

54. See, for example, Stephen Braun, *The Science of Happiness: Unlocking the Mysteries of Mind* (John Wiley, 2000); Robert Cloninger, *Feeling Good: The Science of Happiness* (Oxford University Press, 2004); Eid and Larsen, eds, *Science of Subjective Well-Being*; Huppert, Baylis, and Keverne, eds, *The Science of Well-Being*; Stefan Klein, *The Science of Happiness* (Marlowe, 2006); Daniel Nettle, *Happiness: The Science behind your Smile* (Oxford University Press, 2005). See also, for titles somewhat differently worded, Gregory Berns, *Satisfaction: The Science of Finding True Fulfillment* (Henry Holt, 2005); Richard Layard,

Happiness: Lessons from a New Science (Penguin Press, 2005); Eduardo Punset, *The Happiness Trip: A Scientific Journey* (Chelsea Green, 2007).

55. See Darrin M. McMahon, *Happiness: A History* (Atlantic Monthly, 2006) 212–15.

56. Abbé Pluquet, *De la Sociabilité* (Chez Barrois, 1767). This book is now regarded as one of the forerunners of sociology; earlier, in 1745, Pluquet had published an alphabetically arranged dictionary of heresies. For a discussion of the development, during the eighteenth and nineteenth centuries, of expressions such as "moral sciences" and "human sciences" in France, and the role of forerunners such as Pluquet, see Johan Heilbron, *The Rise of Social Theory*, tr. Sheila Gogol (University of Minnesota Press, 1995).

57. *Encyclopédie, ou dictionnaire raisonné des sciences, des arts, et des métiers* (Paris, 1751–80). See Jean Le Rond d'Alembert, *Preliminary Discourse to the Encyclopedia of Diderot* (1751), tr. and ed. Richard N. Schwab (University of Chicago Press, 1995); Robert Darnton, *The Business of Enlightenment: A Publishing History of the Encyclopédie 1755–1800* (Harvard University Press, 1979).

58. Kwame Anthony Appiah, *Experiments in Ethics* (Harvard University Press, 2008); Owen Flanagan, *The Really Hard Problem: Meaning in a Material World* (MIT Press, 2007), 3; and Daniel M. Haybron, *The Pursuit of Unhappiness: The Elusive Psychology of Well-Being* (Oxford University Press, 2008); Martha Nussbaum and Amartya Sen, *The Quality of Life* (Clarendon Press, 1993); Edward Slingerland, *What Sciences Offer the Humanities: Integrating Body and Culture* (Cambridge University Press, 2006); and Valerie Tiberius, *The Reflective Life: Living Wisely with Our Limits* (Oxford University Press, 2008).

59. William James, *Varieties of Religious Experience*, ed. Fredson Bowers, (Harvard University Press, 1985), 115.

VI. Beyond Temperament

1. William James, *Varieties of Religious Experience*, ed. Fredson Bowers (Harvard University Press, 1985), 115.

2. See Ed Diener and Richard E. Lucas, "Personality and Social Well-Being," in Daniel Kahneman, Ed Diener, and N. Schwartz, eds, *Well-Being: The Foundations of Hedonic Psychology* (Russell Sage Foundation, 1999, 213–29; and Daniel Nettle, *Happiness: The Science behind your Smile* (Oxford University Press, 2005), 91–114.

3. *Oxford English Dictionary*: The word "temperament" derives from the Latin *temperamentum*, due mixture. In medieval physiology, people's physical and mental constitution was held to be influenced by the proportions in which the four cardinal humors of the body interacted.

4. Robert Burton, *The Anatomy of Melancholy*, ed. Holbrook Jackson (New York Review of Books, 2001), 20. See references to Democritus's work "Cheerfulness or Well-Being" in Kathleen Freeman, *Ancilla to the Pre-Socratic Philosophers* (Harvard University Press, 1948), 91 and 101.

5. Ibid., third and fourth stanzas in "The Author's Abstract of Melancholy," 11.

6. Aristotle, *Problems XXX*, 1, in W. S. Hett, tr., *Aristotle Problems* (Loeb Classical Library, Harvard University Press, 1965), vol. II, 154–5. The authenticity of this text has long been questioned. See Jennifer Radden, *The Nature of Melancholy from Aristotle to Kristeva* (Oxford University Press, 2002), 55; and M. A. Screech, *Montaigne and Melancholy: The Wisdom of the Essays* (Duckworth, 1983), 27.

7. Cicero, *Tusculan Disputations*, I, 33.80, tr. J. E. King (Loeb Classical Library, Harvard University Press, 1971), 94.

8. See Screech, *Montaigne and Melancholy*, 28; Seneca, "On Tranquility of Mind," tr. John W. Basore (Loeb Classical Library, Harvard University Press, 1972), 285; Burton, *Anatomy*, 401.

9. Burton, *Anatomy*, 401. Peter Quennell, in *The Pursuit of Happiness* (Constable, 1988), points out that "Malcontents" were often portrayed, as on the frontispiece of Burton's *Anatomy*, wearing black clothes with their "melancholy hats" pulled down toward their eyes and their right hand fingering their dagger. For a discussion of Burton's *Anatomy* and of commentators over the centuries since he wrote, see Charles Rosen, "The Anatomy Lesson," *New York Review of Books*, June 9, 2005, 55–9.

10. Immanuel Kant, *Observations on the Feeling of the Beautiful and the Sublime*, tr. John T. Goldthwait (University of California Press, 1991), 62–70. See also Kant's *Anthropology from a Practical Point of View*, tr. Victor Lyle Dowdell (Southern Illinois University Press, 1978), 198–202.

11. Kant, *Observations*, 66.

12. Ibid., 67.

13. Arthur Schopenhauer, *The World as Will and Representation*, tr. E. F. J. Payne (Dover Publications, 1969), vol. II, 573. See also Eduard von Hartmann's *The Philosophy of the Unconscious*, tr. Coupland, 3 vols (London, 1884). Von Hartmann drew on Schopenhauer and on neurology, psychology, and the history of culture to conclude that there is an "optimistic illusion" based on the illusory belief that happiness is attainable in this world.

14. Arthur Schopenhauer, "On the Wisdom of Life: Aphorisms," in Arthur Schopenhauer, *Complete Essays*, tr. T. Bailey Saunders (Wiley, 1942), 78–9.

15. See, for example, Sonja Lyubomirsky, *The How of Happiness: A Scientific Approach to Getting the Life You Want* (Penguin Press, 2008), 242–3; Ed Diener and Robert Biswas-Diener, *Happiness: Unlocking the Mysteries of Psychological Wealth* (Blackwell, 2008), 74.

16. "Arthur's Advice: Comparing Arthur Schopenhauer's Advice on Happiness with Contemporary Research," *Journal of Happiness Studies*, vol. 9, 2008, 379–95.

17. Arthur Schopenhauer, "On Women," in *Studies of Pessimism*, sel. and tr. T. Bailey Saunders (G. Allen & Unwin, 1923), 106–7.

18. Bertrand Russell, *Wisdom of the West* (Doubleday, 1959), 257.

19. Andrew Solomon, *The Noonday Demon: An Atlas of Depression* (Simon & Schuster, 2001), 78. For accounts of personal experiences with antidepressants, see Mark Kingwell, *Better Living: In Pursuit of Happiness from Plato to Prozac* (Viking, 1998); and Stephen Braun, *The Science of Happiness: Unlocking the Mysteries of Mood* (John Wiley, 2000), Epilogue, 161–81.

20. Among the growing proportions of troops afflicted with post-traumatic stress disorder after returning from modern combat, as also among victims of family violence, what is often at issue is a long period of traumatic experiences rather than some one precipitating event. I discuss this condition and the role of exposure to violence in *Mayhem: Violence as Public Entertainment* (Perseus Books, 1998), 61–6.

21. Quoted in review by Joyce Carol Oates of *The Unabridged Journals of Sylvia Plath, 1950–1962*, ed. Karen V. Kukil (Anchor Books, 2000), in the *New York Times Book Review*, November 5, 2000, 10.

22. One, Propranolol, can ease the pain of such memories, but only if it is administered soon after the traumatic experience. And concerns would mount if

drugs were to be found capable of wiping away painful memories more completely. Might people not suffering from post-traumatic stress disorder also wish to take such drugs to do away with memories of harm they have done to others? Might drugs that induce amnesia diminish the capacity to feel guilt, remorse, and empathy? As David Wasserman suggests, inducing such amnesia may reduce our affinity with our past selves, something that would be "a lot like reducing our empathy for other people." See David Wasserman, "Making Memory Lose Its Sting," *Philosophy & Public Policy Quarterly*, vol. 24, no. 4, 2004, 18.

23. Gro Harlem Brundtland, "Message from the Director-General," in *The World Health Report 2001: Mental Health: New Understanding, New Hope* (World Health Organization, 2001), x.

24. Sir Thomas Browne, *Religio Medici and Other Writings*, ed. M. R. Ridley (Dent, 1965), 65, 83.

25. John Morreall, "Comedy," in Edward Craig, ed., *Routledge Encyclopedia of Philosophy* (Routledge, 1998), vol. II, 434–8.

26. Leslie Stephen, "Sir Thomas Browne," in *Hours in a Library* (G. P. Putnam's Sons, 1904), vol. I, 269–70.

27. Pierre Teilhard de Chardin, *Sur le bonheur*, tr. René Hague, *On Happiness* (Harper & Row, 1973), 10.

28. Ibid.

29. Ursula King, ed., *Pierre Teilhard de Chardin: Essential Writings* (Orbis Books, 1999), 128.

30. Teilhard, *On Happiness*, 44–5.

31. Felicia A. Huppert, "Positive Mental Health in Individuals and Populations," in Felicia A. Huppert, Nick Baylis, and Barry Keverne, eds, *The Science of Well-Being* (Oxford University Press, 2007), 307.

32. Cited by Stephen, "Sir Thomas Browne," 267.

33. See Jonathan Weiner, *Time, Love, Memory* (A. A. Knopf, 1999), 200–11, asking whether inferences for human beings can be drawn from research on inhibiting such switches among fruit flies.

34. In one of his last works, Sir Thomas spoke of "oblivion" – a capacity allowing most people to remain even-tempered despite whatever grievous shocks fate has in store for them: "To be ignorant of evils to come, and forgetfull of evils past, is a merciful provision in nature . . . our delivered senses not relapsing into cutting remembrances, our sorrows are not kept raw by the edge of repetitions." Browne, "Urn Burial," in *Religio Medici*, 136.

35. James, *Varieties of Religious Experience*, 75–9.

36. *Mencius*, Book I, Part A, 6, tr. D. C Lau (Penguin Books, 1976), 82–3.

37. Immanuel Kant, *Metaphysics of Morals*, ed. Mary Gregor, in *Practical Philosophy* (Cambridge University Press, 1996), 528. No more than Mencius did Kant view such feelings to be the only indispensable predispositions; he mentioned, as well, conscience and respect for oneself or self-esteem.

38. The erosion of one can debilitate the other: those who become more fearful, after witnessing or enduring violence, can learn to shield themselves against feeling the pain of others, sometimes by taking vicarious or active pleasure in maiming and killing – exhibiting "learned pitilessness." See my "Violence, Free Speech, and the Media," in *Perspectives on Crime and Justice: 1998–1999 Lecture Series* (National Institute of Justice, 1999), 51–72.

39. Aldous Huxley, *Time Must Have a Stop* (Harper & Brothers, 1944), 3.

40. George L. Vaillant, "Adaptive Mental Mechanisms," *American Psychologist*, January 2000, 89–98, at 90.
41. Immanuel Kant, *Lectures on Ethics*, tr. Peter Heath (Cambridge University Press, 1997), 87.
42. Philip Zimbardo and John Boyd, *The Time Paradox: The New Psychology of Time That Will Change Your Life* (Free Press, 2008), 6.
43. Robert Mauzi, *Maintenant sur ma route . . .* (Paradigme, 1994), 19. See also Robert Mauzi, *L'Idée du bonheur au xviiie siècle* (Librairie Armand Colin, 1960), 254–5.
44. John T. Cacioppo and William Patrick, *Loneliness: Human Nature and the Need for Social Connection* (Norton, 2008).
45. Ed Diener and Robert Biswas-Diener, *Happiness: Unlocking the Mysteries of Psychological Wealth* (Blackwell, 2008), 50.
46. Petrarch, *The Life of Solitude by Francis Petrarch*, tr. Jacob Zeitlin (University of Illinois Press, 1924), 107–8. See Armando Maggi, "You Will Be My Solitude. Solitude as Prophecy. *De vita solitaria*," in Victoria Kirkham and Armando Maggi, eds, *Petrarch: A Critical Guide to the Complete Works* (University of Chicago Press, 2009), 179–95; and Michael O'Loughlin, *The Garlands of Repose: The Literary Celebration of Civic and Retired Leisure* (University of Chicago Press, 1978), 227–32.
47. Petrarch, *Life of Solitude*, 292.
48. O'Loughlin, *Garlands of Repose*, 232, citing Zeitlin's translation of *De vita solitaria* and the Latin original.
49. Rabindranath Tagore, *Letters to a Friend*, ed. C. F. Andrews (Macmillan, 1929), 22.
50. Quennell, *The Pursuit of Happiness*, 188.
51. Ibid., 80.
52. Letter to William Howard Schubert, July 25, 1952, in Jack Cowart and Juan Hamilton, eds, *Georgia O'Keeffe: Art and Letters* (Little, Brown, 1987), 263.
53. Winslow Homer, letter, 1895, Clark Museum, Williamstown, Massachusetts.
54. Peter Quennell, *The Marble Foot: An Autobiography 1905–1938* (Viking Press, 1976), 245.
55. Ernst Samuels, *Bernard Berenson: The Making of a Legend* (Harvard University Press, 1987), 1.
56. Bernard Berenson, *Sunset and Twilight: From the Diaries of 1947–1958*, ed. Nicky Mariano, (Harcourt, Brace & World, 1963), 135, entry for July 3, 1949.
57. Iris Origo, *Images and Shadows: Part of a Life* (David A. Godine, 1970), 132–3.
58. Alexander Nehamas, *Only a Promise of Happiness* (Princeton University Press, 2007), 138.
59. L. W. Sumner, *Welfare, Happiness and Ethics* (Clarendon Press, 1996), 20–2.
60. Ibid., 8–11.

VII. Is Lasting Happiness Achievable?

1. Sigmund Freud, *Civilization and its Discontents*, tr. and ed. James Strachey, with an Introduction by Louis Menand and a Biographical Afterword by Peter Gay (W. W. Norton, 2005), 53.
2. Bertrand Russell, *The Conquest of Happiness* (W. W. Norton, 1996), 11.
3. Freud, *Civilization*: "Our enquiry concerning happiness [*Unsere Untersuchung über das Glück*] has not so far taught us much that is not already common knowledge," 67. Freud's letter to Lou Andreas-Salomé, written on July 28, 1929, the day he finished writing *Civilization*, is quoted in the "Einleitung" to *Das Unbehagen in der Kultur* (Fischer Taschenbuch, 2000), 12.

4. Russell, *Conquest*, 11.
5. Louis Menand, Introduction to Freud, *Civilization*, 9.
6. Russell, who admired Freud as the man who had revolutionized psychology in their century and dared to speak of childhood sexuality, had signed a petition the same year their books appeared, urging that Freud be awarded the Nobel Prize in medicine.
7. Russell, *Conquest*, 17; Freud, *Civilization*, 119.
8. Freud, *Civilization*, 53.
9. Ibid., 53–4. (Strachey translated Freud's "andere Menschen" as "other men.")
10. Russell, *Conquest*, 130.
11. Ibid., 191.
12. See, among Bertrand Russell's early writings on such topics, his *Principles of Social Reconstruction*, tenth reprinting (Unwin Brothers, 1930).
13. Freud, *Civilization*, 62.
14. Ibid., 58.
15. Ibid., 104.
16. Sigmund Freud, *The Future of an Illusion* (W. W. Norton, 1987), 2.
17. Letter to Oskar Pfister, February 7, 1930, quoted in Peter Gay, *Freud: A Life for Our Time* (W. W. Norton, 1988), 552. See 523–53 for Gay's discussion of *Civilization and its Discontents*.
18. Ibid., 525.
19. Freud's letter to Lou Andreas-Salomé July 28, 1929. (See note 3, above.)
20. Freud, *Civilization*, 40. Freud, who rejected the use of contraceptives in his own marriage, added: "What is the use of reducing infantile mortality when it is precisely that reduction which imposes the greatest constraint on us in the begetting of children, so that, taken all round, we nevertheless rear no more children than in the days before the reign of hygiene, while at the same time we have created difficult conditions for our sexual life in marriage, and have probably worked against the beneficial effects of natural selection?"
21. Bertrand Russell, "A Free Man's Worship," in Paul Edwards, ed., *Why I Am Not a Christian and Other Essays* (Simon & Schuster, 1957), 113.
22. Russell, *Conquest*, 123.
23. *The Autobiography of Bertrand Russell, 1914–1944* (Little, Brown, 1968), 230.
24. Dora Russell, *The Right to Be Happy* (Harper & Brothers, 1927), 295.
25. Russell, *Autobiography, 1914–1944*, 238–9.
26. Russell, *Conquest*, 151.
27. Katharine Tait, *My Father Bertrand Russell* (1975; Thoemmes Press, 1996), 202.
28. See, for contrasting views on Russell's sincerity, Ronald Clark, *The Life of Bertrand Russell* (Alfred A. Knopf, 1976), 448; and Ray Monk, *Bertrand Russell: The Ghost of Madness* (Free Press, 2000), 114.
29. Ruut Veenhoven, "Happiness as an Aim in Public Policy," in Alex Linley and Stephen Joseph, eds, *Positive Psychology in Practice* (John Wiley, 2004), 8.
30. J. M. Roberts, *The Twentieth Century: The History of the World, 1901–2000* (Viking Press, 1999), 837.
31. Freud, *Civilization*, 60.
32. Slavoj Žižek, *Welcome to the Desert of the Real* (Verso, 2002), 60; Jacques Lacan, Seminar XVII, in Jacques Lacan, *Ecrits: A Selection* (Routledge, 2001).
33. Grant Duncan, "After Happiness," *Journal of Political Ideologies*, vol. 85, 2007, 86–108.

34. Philip Brickman and Donald T. Campbell, "Hedonic Relativism and Planning the Good Society," in M. H. Apley, ed., *Adaptation Level Theory: A Symposium* (Academic Press, 1971), 287–302.

35. Oliver Goldsmith, *The Citizen of the World or Letters from a Chinese Philosopher, Residing in London, to His Friends in the East* (1762), in *Collected Works*, ed. Arthur Friedman (Clarendon Press, 1966), vol. II, 185–6.

36. Ed Diener, E. M. Suh, Richard E. Lucas, and Heidi L. Smith, "Subjective Well-Being: Three Decades of Progress," *Psychological Bulletin*, vol. 125, no. 2, 1999, 276–302, at 285; Shane Frederick and George Lowenstein, "Hedonic Adaptation," in Daniel Kahneman, Ed Diener, and Norbert Schwartz, eds, *Well-Being: The Foundations of Hedonic Psychology* (Russell Sage Foundation, 1999), 302–29. See also Stephen Braun, *The Science of Happiness: Unlocking the Mysteries of Mind* (John Wiley, 2000); and Richard A. Easterlin, "The Economics of Happiness," *Daedalus*, Spring 2004, 28.

37. Daniel T. Gilbert, Elizabeth C. Pinel, Timothy D. Wilson, Stephen J. Blumberg, and Thalia P. Wheatley, "Immune Neglect: A Source of Durability Bias in Affective Forecasting," *Journal of Personality and Social Psychology*, vol. 75, September 1998, 617–38.

38. See, for a list of "elements of adaptation", Paul Menzel, Paul Dolan, Jeff Richardson, and Jan Abel Olsen, "The Role of Adaptation to Disability and Disease in Health State Valuation: A Preliminary Normative Analysis," *Social Science & Medicine*, vol. 55, 2002, 2149–58.

39. Richard A. Easterlin, "Subjective Well-Being and Economic Analysis: A Brief Introduction," *Journal of Economic Behavior and Organization*, vol. 45, 2001, 225–6.

40. Ed Diener and Robert Biswas-Diener, *Happiness: Unlocking the Mysteries of Psychological Wealth* (Blackwell, 2008), 91, 110.

41. Ed Diener, Richard E. Lucas, and Christie Napa Scollon, "Beyond the Hedonic Treadmill: Revising the Adaptation Theory of Well-Being," *American Psychologist*, vol. 61, May–June 2006, 305–14, at 303.

42. David Lykken and Auke Tellegen, "Happiness is a Stochastic Phenomenon," *Psychological Science*, vol. 7, May 1996, 186–9.

43. Ibid., 189.

44. David Lykken, *Happiness: The Nature and Nurture of Joy and Contentment* (St Martin's Griffin, 1999), 2.

45. Ibid., 62.

46. Easterlin, "Economics of Happiness," 27. See also Erik Parens, "Genetic Differences and Human Identities: On Why Talking about Behavioral Genetics Is Important and Difficult," *Hastings Center Report*, January–February 2004, Special Supplement, S4–S35; and Steven Pinker, "Why Nature and Nurture Won't Go Away," *Daedalus*, Fall 2004, 5–17.

47. Bruce Headley, "The Set-Point Theory of Well-Being Needs Replacing – On the Brink of a Scientific Revolution?" October 2007, Social Science Electronic Publishing, Inc.

48. Ronald F. Inglehart, Roberto Foa, Christopher Peterson, and Christian Welzel, "Development, Freedom, and Rising Happiness: A Global Perspective," *Perspectives on Psychological Science*, vol. 3, July 2008, 264–85, at 264.

49. Felicia A. Huppert, "Positive Mental Health in Individuals and Populations," in Felicia A. Huppert, Nick Baylis, and Barry Keverne, eds, *The Science of Well-Being* (Oxford University Press, 2007), 317.

50. Matt Ridley, *Nature via Nurture: Genes, Experience, and What Makes Us Human* (HarperCollins, 2003), 6.

51. Arthur Schopenhauer, *The World as Will and Representation*, tr. E. F. J. Payne (Dover Publications, 1958), vol. II, 573.

52. John Stuart Mill, *Autobiography* (1873; Oxford University Press, 1971), 101.

53. Ibid., 102.

54. Martin P. Seligman, *Authentic Happiness: Using the New Positive Psychology to Realize Your Potential for Lasting Fulfillment* (Free Press, 2002), 45. See also Tal Ben-Shahar, *Happier: Learn the Secrets to Daily Joy and Lasting Fulfillment* (McGraw Hill, 2007)

55. Sonja Lyubomirsky, *The How of Happiness: A Scientific Approach to Getting the Life You Want* (Penguin Press, 2008).

56. Diener and Biswas-Diener, *Happiness*, 3.

57. Ibid., 6.

58. See Eric Weiner's amusing commentaries in *The Geography of Bliss: One Grump's Search for the Happiest Places in the World* (Twelve, 2008).

59. Robert A. Emmons, "Gratitude, Subjective Well-Being, and the Brain," in Michael Eid and Randy J. Larsen, eds, *The Science of Well-Being* (Guilford Press, 2008), 169–89.

60. Margaret Visser, *The Gift of Thanks* (HarperCollins, 2008).

61. Ibid., 173.

62. Diogenes Laertius, *Lives of Eminent Philosophers*, tr. R. D. Hicks (Loeb Classical Library, Harvard University Press, 1972), vol. I, 35.

63. Simone Weil, "Draft for a Statement on Human Rights," in *Two Moral Essays* (Pendle Hill, 1981), 13.

64. Simone Weil, "On Personality," ibid., 28.

VIII. Illusion

1. Augustine, *Confessions*, Book X, 22, tr. R. S. Pine-Coffin (Penguin Books, 1961), 228.

2. Karl Marx, "Toward a Critique of Hegel's Philosophy of Right," in David McClellan, *Karl Marx: Selected Writings* (Oxford University Press, 2000), 72.

3. Abu Hamid al-Ghazali, *The Alchemy of Happiness*, tr. Claud Field (M. E. Sharpe, 1991), ch. VIII, 107–8. For discussion, see Ignacio L. Götz, *Conceptions of Happiness* (University Press of America, 1995), 217–44.

4. Karl Marx, "Critique of the Gotha Programme," in McClellan, ed., *Karl Marx: Selected Writings*, 615.

5. Aquinas, *Summa Theologica*, tr. Fathers of the English Dominican Province (Christian Classics, 1981), First Part of the Second Part, Question 3, Article 8.

6. Later Buddhists of different schools may offer more specifics about nirvana but still see it as entirely different from the vision of a personal God.

7. See documents quoted in "Buddhism," in Sarvepalli Radhakrishnan and Charles A. Moore, *A Sourcebook in Indian Philosophy* (Princeton University Press, 1973), 272–346. For contemporary Buddhist discussions of happiness and illusion, see B. Alan Wallace, *Buddhism with an Attitude* (Snow Lion Publications, 2001); Dalai Lama, *Stages of Meditation* (Snow Lion Publications, 2001); Similar doctrines were mentioned in the Bhagavadgita and in the Dialogues of the Buddha.

8. Radhakrishnan and Moore, "Cārvāka" in *Sourcebook in Indian Philosophy*, 227–8.

9. *The Analects of Confucius*, tr. Simon Leys (W. W. Norton, 1997), 50.

10. Horace, Epistle II. 2. 138–40, in *The Complete Works of Horace*, tr. Charles E. Passage (F. Ungar, 1986), 354. See also translation of this Epistle by Smith Palmer Bowie, in *Satires and Epistles of Horace* (University of Chicago Press, 1959), 266. All the translations of this Epistle that I have seen use the term "illusion" for Horace's "error," which has meanings ranging from "wandering" to "error," "mistake," "illusion" and "delusion." Horace compared the false bliss of Lycas to the illusions that poets who write poor verse maintain about their talent – a type infinitely more common to readers than the exotic figure of Lycas.

11. Erasmus, *The Praise of Folly and Other Writings*, tr. and ed. Robert M. Adams (W.W. Norton, 1989), 39.

12. Montaigne, *The Complete Essays of Montaigne*, tr. Donald M. Frame (Stanford University Press, 1965), "Apology for Raymond Sebond," 365–6.

13. Alexander Pope, "Imitations of Horace," II, ii, in *The Poems of Alexander Pope*, ed. John Butt (Yale University Press, 1963), 655.

14. Robert Mauzi, *L'Idée du bonheur au xviiie siècle* (Librairie Armand Colin, 1960), 16–18.

15. Francis Bacon, "Of Truth," in *Essays* (J. M. Dent, 1972), 3.

16. Julien Offray de La Mettrie, *De la Volupté: Anti-Sénèque ou le souverain bien; L'Ecole de la volupté, Système d'Epicure* ed. Ann Thomson (Les Editions Des Jonquières, 1996), 36.

17. Madame du Châtelet, *Discours sur le bonheur* (Rivage Poche Petite Bibliothèque, 1997), 32 [my translation]. See also Elisabeth Badinter, *Emilie, Emilie: L'Ambition féminine au XVIIIème siecle* (Flammarion, 1983).

18. Du Châtelet, *Discours sur le bonheur*, 66. When Madame du Châtelet wrote her reflections on happiness, she had not yet engaged in her second great love affair, with the Marquis de Saint-Lambert, ten years her junior. She became pregnant, at forty-three, and died after giving birth to a daughter, who died shortly thereafter.

19. Mauzi, *L'Idée du bonheur*, 9.

20. For an account of her last months, the baby's birth, and her death, see Badinter, *Emilie, Emilie*, 451–68.

21. Ibid., 431.

22. See James M. Bradburne, "More than Meets the Eye," in *Amazing Illusions: The Wonders of Trompe l'oeil* (Alias, 2009); and Ernst Gombrich, *Art and Illusion* (Princeton University Press, 1969), "Conditions of Illusion," 203–41.

23. Shelley E. Taylor, "Professional Profile." http://shelley.taylor.socialpsychology.org.

24. See Shelley E. Taylor, *Positive Illusions: Creative Self-Deception and the Healthy Mind* (Basic Books, 1989); Shelley E. Taylor and J.D. Brown, "Illusion and Well-Being: A Social Psychological Perspective on Mental Health," *Psychological Bulletin*, vol. 103, 1988, 193–210; and Shelley E. Taylor, Margaret E. Kemeny, Geoffrey M. Reed, Julienne E. Bower, and Tara L. Grunewald, "Psychological Resources, Positive Illusions, and Health," *American Psychologist*, January 2000, 99–109.

25. Shelley E. Taylor, "On Healthy Illusions," *Daedalus*, Winter 2005, 134–5.

26. See C. Randall Colvin and Jack Block, "Do Positive Illusions Foster Mental Health? An Examination of the Taylor and Brown Formulation," *Psychological Bulletin*, vol. 116 1994, 3–20; and C. Randall Colvin and Jack Block, "Positive Illusions and Well-Being Revisited: Separating Fact from Fiction," ibid., 21–7.

27. See the chapter on "Self-Deception," in my book *Secrets: On the Ethics of Concealment and Revelation* (Pantheon Books, 1983), especially 69–71.

28. Taylor, *Positive Illusions*, 126.

29. Harry Stack Sullivan, *The Interpersonal Theory of Psychiatry* (Norton, 1953), 319.

30. Frank R. Sloan, V. Kerry Smith, and Donald H. Taylor, Jr., *The Smoking Puzzle: Information, Risk Perception, and Choice* (Harvard University Press, 2003).

31. Martin E. P. Seligman, *Authentic Happiness: Using the New Positive Psychology to Realize your Potential for Lasting Fulfillment* (Free Press, 2002), 200; see also Sandra Murray et al., "The Quest for Conviction: Motivated Cognition in Romantic Relationships," *Psychological Inquiry*, (1999) vol. 10, 23–34.

32. Seligman, *Authentic Happiness*, 202.

33. Badinter, *Emilie, Emilie*, 132, 427.

34. Bertrand Russell, *Sceptical Essays* (Routledge Classics, 2004), 16.

35. Daniel Gilbert, *Stumbling on Happiness* (Alfred A. Knopf, 2006). See also George Loewenstein and David Schkade, "Wouldn't It Be Nice? Predicting Future Feelings," in Daniel Kahneman et al., *Well-Being: The Foundations of Hedonic Psychology* (Russell Sage Foundation, 1999), 84–105.

36. Jonathan Swift, "A Digression on Madness," in A. C. Guthkelch and D. Nichol Smith, eds, *A Tale of a Tub* (Clarendon Press, 1958), 171. According to a note on this definition of happiness, by the editors, "Swift is thought by Sir Henry Craik to have had in his mind the passage in Horace (*Epistle*. II. ii. 140)."

37. Ibid., 174.

38. Eric R. Kandel, *In Search of Memory: The Emergence of a New Science of Mind* (W. W. Norton, 2006), 450.

39. Ibid., 311, quoting William James, *Principles of Psychology*; and 314.

40. Barbara Herman, *The Practice of Moral Judgment* (Harvard University Press, 1993), 77–93.

41. Iris Murdoch, *The Sovereignty of Good* (Penguin, 1992), 59.

42. Murdoch, *Metaphysics as a Guide to Morals* (Routledge & Kegan Paul, 1985), 249–50.

43. Barbara Ley Toffler, with Jennifer Reingold, *Final Accounting: Ambition, Greed, and the Fall of Arthur Andersen* (Broadway Books, 2003).

44. Sissela Bok, "Impaired Physicians: What Should Patients Know?" *Cambridge Quarterly of Healthcare Ethics*, vol. 2, 1993, 331–40.

45. Joyce Ehrlinger, Kerry Johnson, Matthew Banner, David Dunning, and Justin Kruger, "Why the Unskilled Are Unaware: Further Exploration of (Absent) Self-Insight among the Incompetent," *Organizational Behavior and Human Decision Processes*, vol. 105, 2008, 98–121, at 98. See also Justin Kruger and David Dunning, "Unskilled and Unaware of it: How Difficulties in Recognizing one's own Incompetence Lead to Inflated Self-Assessment," *Journal of Personality and Social Psychology*, vol. 77, no. 6, 1999, 1121–34.

46. James Griffin, *Well-Being: Its Meaning, Measurement, and Moral Importance* (Cambridge University Press, 1993), 9.

47. Wayne Sumner, "Something in Between," in Roger Crisp and Brad Hooker, eds, *Well-being and Morality: Essays in Honour of James Griffin* (Oxford University Press, 2000), 5–6. See also Kwame Anthony Appiah, *Experimental Ethics* (Harvard University Press, 2008), ch. 5; Robert Nozick, *Socratic Puzzles* (Harvard University Press, 1997); and Amartya Sen, *The Idea of Justice* (Harvard University Press, 2009), ch. 7.

48. Immanuel Kant, *Anthropology from a Practical Point of View* (1798; tr. Victor Lyle Dowdell, Southern Illinois University Press, 1978), 207. I discuss this passage in "Shading the Truth in Seeking Informed Consent for Research Purposes," *Kennedy Institute of Ethics Journal*, vol. 5, 1995, 1–17; and in "Truthfulness," in Edward Craig, ed., *Routledge Encyclopedia of Philosophy* (Routledge, 1998), 480–5, at 484.

49. *Mencius*, tr. D. C. Lau (Penguin Books, 1976), 182.

IX. The Scope of Happiness

1. The *Oxford Compact Thesaurus* gives as examples of the first meaning "the scope of the investigation," and, of the second, "the scope for change is limited by political realities." See "Scope," in *Oxford Compact Thesaurus* (Oxford University Press, 2001), 760.

2. Ivor A. Richards, *Principles of Literary Criticism* (Kegan, Paul, Trench, Trübner, 1925), 40. See Chapter 2, above.

3. In medical research, the term is also used when a trial using human subjects is brought to a close earlier than planned, for safety reasons.

4. Sarah Broadie speaks of "the scope of happiness" in discussing Aristotle's treatment of how health and prosperity relate to *eudaimonia*, "On the Other Goods and the Scope of Happiness," in *Ethics with Aristotle* (Oxford University Press, 1991), 50.

5. Willa Cather, inscription on gravestone in Jaffrey, Vermont; Betty Sue Flowers, ed., *Joseph Campbell and the Power of Myth* (Doubleday, 1988), 113.

6. Jonathan Swift, "A Digression on Madness," in A. C. Guthkelch and D. Nichol Smith, eds, *A Tale of a Tub* (Clarendon Press, 1958), 171.

7. Sissela Bok, *Lying: Moral Choice in Public and Private Life*, 2nd edn (Vintage Books, 1999); *Secrets: On the Ethics of Concealment and Revelation*, 2nd edn (Vintage Books, 1989); *Mayhem: Violence as Public Entertainment* (Perseus Books, 1998); and *Common Values*, 2nd edn (University of Missouri Press, 2002). The following sentences draw on the Preface to the 2002 edition of *Common Values*.

8. Michel de Montaigne, "Of Repentance," tr. Donald Frame, in *The Complete Essays of Montaigne* (Stanford University Press, 1965), 611. I have substituted, in Frame's translation, "the human condition" in rendering Montaigne's "l'humaine condition," instead of "man's estate."

INDEX

ACKNOWLEDGMENTS

I feel fortunate to have had a chance to discuss questions of happiness with so many friends, family members, and colleagues while working on this book. For valuable advice, I want especially to thank Daniel Callahan, Sidney Callahan, Lincoln Chen, Martha Chen, Sara Esgate, Howard Gardner, Ellen Goodman, Neva Goodwin, Martha Nussbaum, the late Robert Nozick, Amartya Sen, Brita Stendahl, and Susan Tifft. I am grateful to have been able to take part in an academic working group on happiness, under the auspices of the Harvard Mind Brain Behavior Initiative, that included Daniel Gilbert, Allan Brandt, Nancy Etcoff, and David Leibson. I am greatly indebted to those who were kind enough to comment on parts of the manuscript: Ann Hunter, John Martin, Richard Newhauser, Emma Rothschild, and Richard Watson. My warm thanks go to my editor Heather McCallum at Yale University Press, for her support for the book from the outset and for indispensable guidance and most helpful and imaginative suggestions; to Beth Humphries for her meticulous copy-editing, and to Tami Halliday and Rachael Lonsdale for all their help during the later stages of production.

To my husband Derek, finally, goes my deepest gratitude. As with my earlier books, his encouragement, interest, and careful scrutiny of each chapter were invaluable. But this time around, it was my great good fortune that he was also writing a book on happiness – *The Politics of Happiness*. For the two of us to have worked on books on the same subject from such different perspectives, discussing texts and ideas, reading and critiquing drafts, has been a source of happiness in its own right.